EMPTY PLATES

What people are saying about *EMPTY PLATES*

"Meeting Dave Krepcho a few years ago during a visit to Orlando's Second Harvest Food Bank, I was immediately impressed with his leadership and dedication to feeding the hungry. As I came to know him better, I was enraptured by his innovative approach to the challenges of hunger relief and his storytelling ability. This is a seminal work on the national food bank story, and it's something all Americans should read...as it will help us all to contribute."

- Mark Hertling, Lieutenant General (retired)

"As a pastor in Central Florida, I watched Dave Krepcho's remarkable leadership bring a good part of our state together to serve those with empty plates. Food is about more than physical nutrition; it is about people's stories - both those in need and those who want to help. This book not only educates and fascinates us, it expands our understanding, our hearts, and our desire for partnership."

- Dr. Joel C. Hunter, Pastor of Community Benefit, Action Church

"Empty Plates by Dave Krepcho pays attention to the periphery of Food Banking as only he can. He has spent 34 years of his life understanding Food Banking through authentic empathy, genuine compassion, and creative leadership."

- Stanley S. Gryskiewicz, Ph.D., A member of the founding staff of the Center for Creative Leadership (1970) and the founder of the Association of Managers of Innovation in 1980.

What people are saying about *EMPTY PLATES*

"Dave Krepcho reminds us that millions of people in the United States are hungry. Then, he demonstrates what can be done to design hunger out and abundance in. His work evidences that empty plates can be filled with nutritious food and offered up with love. He lights a path to possibility in times that can feel dark and offers real solutions to a moral imperative. He has done the deed. His words resonate. And the leadership lesson is clear-we can and must address inequity as part of our leadership practice."

- Mary McBride, Chair, Creative Enterprise Leadership Development, Pratt

"Dave Krepcho & Claire Strom did something that all of us who worked in food banking hoped one day would be done. To capture the history of hunger relief in America and the evolution of food banking from its very beginning. It tells the story of how a small group of courageous social entrepreneurs built a movement that would change an industry and communities all over America and eventually the world."

- Bill Bolling, Founder, Atlanta Community Food Bank (retired), Food Well Alliance, and Atlanta Regional Housing Forum

"This book is a "Must Read" for anyone who wants to fully understand the American food system and the powerful role that food banks have played in making certain that people experiencing hunger are not forgotten among that system's priorities. Kudos for memorializing the history and the now of this great network."

- Claire Babineaux-Fontenot, CEO, Feeding America

What people are saying about *EMPTY PLATES*

"Being deeply rooted in the culinary world for most of my life, I've witnessed firsthand the transformative power that food holds. The flavors, traditions, and rituals surrounding food creates our family bonds and shapes the very culture of our communities. A must-read for anyone eager to delve into the rich tapestry of culinary heritage and the profound connections it fosters."

- John Rivers, Founder & CEO
4R Restaurant Group, 4Roots Farm

"When I began serving as the Florida Regional Organizer with Bread for the World in 2017, David Krepcho was one of the first allies I made in the work of addressing and ending food insecurity. David held a wealth of information and history on the origin of food banking and shared it freely with me. This knowledge base was foundational to the work that we would embark on together. David opened doors to me that I otherwise would not have been invited to walk into. The concepts of social determinates of health and intersectionality were being spoken by David way before these terms would become an integral part of the framework of our mission to end hunger. David had a keen understanding of how the food system needed to function and where the inefficiencies were. Dave was a visionary, connecting the dots to things like food as medicine, the importance of community collaboration and equitable programs and policies that lead to jobs for people to have the promise of a brighter tomorrow. Dave was never afraid to lend his voice so that others could feel empowered to use theirs."

- Florence French Fagan
Senior Southeast Regional Organizer
Bread For The World

EMPTY PLATES

The evolution of hunger relief
TILL ALL ARE FED

BY
DAVE KREPCHO

With a History of Food Hunger by
CLAIRE STROM

Storytellers Publishing
Colorado, USA

Storytellers Publishing
An imprint of Journey Institute Press,
a division of 50 in 52 Journey, Inc.
journeyinstitutepress.org

Library of Congress Cataloging-in-Publication Data

Names: Krepcho, Dave
Title: EMPTY PLATES
Description: Colorado: Storytellers Publishing, 2025
Identifiers: Library of Congress Control Number: 2024936661
ISBN 979-8-9894379-4-8 (hardcover)
978-1-964754-31-4 (paperback)
Subjects: | BISAC: BUSINESS & ECONOMICS / Industries / Food
Industry. | TECHNOLOGY & ENGINEERING / Food Science /
Food Safety & Security. | BIOGRAPHY & AUTOBIOGRAPHY /
Business.

First Edition
Printed in the United States of America
1 2 3 5 7 19 38 40 72 73

Set in Addington CF / Etna
Editing by Jessica Medberry, InkWhale Editorial LLC.
Design by WiggleB Studios

This book is dedicated to Joseph M. Sciortino
1931–2009
Dear friend and mentor

CONTENTS

Preface

What is not recorded is not remembered.
-Benazir Bhutto

Just before I retired in 2021, after thirty-four years in food banking, it dawned on me that no one had documented the fifty-six-year history (1967–2023) of the national food bank movement in the United States. Capturing it before too much institutional knowledge was lost was important. A hunger relief movement as powerful as food banks deserves to be recorded for various reasons, such as inspiration, understanding, learning, and reflection. History is an explanation of the present. Several pioneers of the food banking movement have passed, and many of us are in the later season of life, so the sooner the better when it comes to documentation. I wanted to pay tribute to those early pioneers and capture some history now, in hopes that later generations will add to it. My passion for the people I've met along the way also prompted me to write this book. It is vital to learn from the past, but not to take it as a perfect map of the space ahead. Hunger relief is an intractable problem—not impossible to solve, but an incredible challenge for many reasons. Everyone deserves access to healthy food; it's a fundamental need. We can end hunger, and we must.

Beyond recording history, my main reason for writing this book is to help change how we think about hunger relief and to drive that change. I hope the book will alter how we innovate, communicate, develop public policy, make decisions, and more. The world around us needs to change in all kinds of ways. As Bob Dylan sang: "The times are a-changin'." One thing is constant: change.

One other reason for writing the book came to me quite unexpectedly and powerfully. Early in the writing process, a few of my grandkids were visiting, and one of them, Raegan, walked into my home office to ask what my book would be about and how far along I was in writing. I said I was just at a rough draft stage. She asked if she could read some of it, so I handed her a chapter, and she stood up and started reading it out loud. That moment hit me hard; it so moved me, and I realized if there was one particular reason I was writing the book, it was for the kids and the future.

Good fortune struck early on when I considered writing this book. I was seeking advice and getting feedback on the concept when Grant Cornwell, president of Rollins College in Winter Park, Florida, introduced me to Claire Strom, Rapetti-Trunzo Chair of History. Over a cup of coffee, Claire informed me of her interest in collaborating on the book and told me her specialty is agriculture. That was one of those "hints from heaven" moments telling me to proceed. Claire and I decided to expand the book's scope to look at hunger relief globally over the centuries. I thought that this broad context would greatly enrich the reading experience. It has been my pleasure to write this book with Claire. I have learned much from her through her editing and her background as a professor and author.

At this point in my life, I want to "pay it forward." So many people I've met have influenced, inspired, taught, mentored, and humored me that it is time to pay back. I have been fortunate to have been surrounded by so many extraordinary people, and I have been able to wire the wisdom of others into my instincts and behavior. This book includes diverse voices ranging from a Lieutenant General in the Army to food bank staff, neighbors facing hunger, volunteers, CEOs, philanthropists, college presidents, elected officials, faith leaders, and more. They have inspired me and reminded me of what's essential. This book is a curated and edited collection, not just a look through my lens at life.

I'll share personal stories of my journey to illustrate my points. Stories can bring us together. I will share moving, enlightening, thought-provoking, educational, and humorous stories. Some of these are very short; I have called them "mission moments" over the years. While writing and interviewing people for this book, I attempted to avoid my personal bias. However, after being inside food banking for so many years, it inevitably found its way into the narrative at times. I hope it does not distort any of the messages.

Although this book is specific to food banks, it can benefit a variety of people: staff within nonprofits, board members, volunteers, elected officials, and financial donors. It can benefit people in corporate or those working with private foundations and government, as well as philanthropic individuals, high school or college students interested in the humanities, and in general, people with curiosity.

Mainly, this book is for those connected and involved with nonprofits, primarily in the human services arena, but it is not necessarily exclusive to that discipline. I have found while coaching at various nonprofits that several topics addressed in the book apply universally, whether in the arts, education, civic engagement, health care, or another field—for example, topics like vision, leadership, and innovation. For nonprofit organizations, this book will be informative for people who are newer in their roles. With the baby boomer generation retiring, many younger people are entering the nonprofit space. Some of this content may already be familiar to seasoned nonprofit executives, but it will serve as a great reminder. Each audience will learn something different. As Maya Angelou said, "No man can know where he is going unless he knows exactly where he has been and exactly how he arrived at his present place."[1]

---

[1] Maya Angelou, "Involvement in Black and White," interview, *Oregonian*, February 17, 1971.

Acknowledgments

I have many people to thank for their contributions. I conducted one hundred interviews with people inside and outside of food banking. No one refused to contribute, for which I am grateful. I especially appreciate the transparency and willingness for conversations that was shown by neighbors facing hunger; their stories are truly inspiring. All the voices within the book have certainly added to its impact. In addition, several people contributed outside of the interview process by providing background information, essential facts, stats, and dates. I appreciate the moral support from several authors who generously offered guidance:

Erica Astazio	Claire Babineaux-Fontenot
David Beckman	Rodney Bivens
Gary Blanchette	Bill Bolling
J. B. Boonstra	Eric Bost
Stephanie Bowman	Mark Brewer
Al Brislain	Karen Broussard
Lori Bruce	Nancy Brumbaugh
Dan Butt	Carrie Calvert
Ken Chapman	Jim Coffin
Patrick Colley	Bill Collins
Kristin Collins	Grant Cornwell
Peg Cornwell	Eric Davis
Lisa Davis	Jaynee Day
Stacy Dean	Patti Delacruz
Kathleen DiChiara	Stephanie Dragatsis
Mike Eagen	Elaine Ellis
Steve Engel	Rob Fersh

EMPTY PLATES

Anterio Fleming
Carol Garrity
Brian Greene
Matt Habash
Mike Halligan
Dawn Haynes
Mark Hertling
Forough Hosseini
Pam Irvine
Kyle Johnson
Jeannie Kiriwas
Martin Krepcho
Linda Landman-Gonzalez
Diane Letson
Lisa Lochridge
Patty Maddox
Thomas Mantz
Bahiyyah Maroon
Jim McGovern
Micki Meyer
Paul Morgan
Lauren Moscowitz
Muhammad Musri
Thao Nguyen
Mindy Ortiz
Denise Parris
Lisa Portelli
Kaye Schmitz
Maria Shanley
John Rivers

Stephanie Garris
Enas Genobli
Stan Gryskiewicz
Vince Hall
Alyssa Harris
Jenny Heaton
Greg Higgerson
Joel Hunter
Barbie Izquierdo
Mike Kazasis
Dawn Koffarnus
Stephanie Krick
Amy Lein
Margaret Linnane
Ricky Ly
Kate Maehr
Santos Maldonado
Mary McBride
Heather McPherson
Eddie Moratin
Susannah Morgan
Carrie Morgridge
Kenny Neal
Maggie O'Halloran
Denise Osterhues
Sabeen Perwaiz
Kelly Quintero
Susan Reimer-Sifford
Erika Spence
Robin Safley

John Sayles	Hilary Seligman
Marsha Semmel	Liz Simeone
Min Sun Kim	Joe Tankersley
June Tanoue	Ellen Teller
Gary Tester	Jeff Thacker
Bonita Thomas	Elaine Waxman
Kirsten White	

Special gratitude goes out to some friends who came before me; a few were instrumental in the startup of the food bank network. Their contribution to food banking history was extremely valuable for this book. Special thanks to Bill Bolling, Rodney Bivens, Al Brislain, Jaynee Day, Kathleen DiChiara, Brian Greene, Matt Habash, and Margaret Linnane. Also, thanks to Carol Garrity, a longtime staffer at Feeding America, for providing various pieces of history about the food bank network. I also appreciate her encouragement at the start of writing this book. There are so many others, too numerous to mention; thank you all.

This book would not have been possible without the incredible support of the Glenda G. Morgan Charitable Foundation. They truly have been special friends and colleagues over the years.

I want to thank my publisher, Michael Jenet of Journey Institute Press, for believing in me and for the contributions of his talented team.

Finally, writing this book has reminded me of being so fortunate in my life at home. I want to thank my wife Lois for her patience, belief in me, and many conversations, comments, and suggestions about the book. Her continual support made this possible.

Introduction

*Life is divided into three terms—that which was,
which is, and which will be. Let us learn from
the past to profit by the present, and from the
present, to live better into the future.*
- William Wordsworth

Anybody in the United States who watched the news during the spring of 2020 saw hundreds or thousands of cars waiting in line to receive food. Sure, COVID-19 was behind it all, but why were those lines so long, so often? Why did the lines for meals at soup kitchens and shelters include more children and seniors than ever before?

These questions lead to others that are perhaps more difficult to answer. Why, over the centuries and around the world, is hunger such an intractable problem? Why isn't more charitable food relief solving the problem? This book will explore how hunger is a symptom of more significant issues.

In the United States, charitable organizations and federal, state, and local governments provide billions of meals annually, but hunger is still present. At this point in history, we have learned enough about the hunger dilemma and can achieve a major turning point in ending it. This book will provide examples of the fight against hunger and detail promising, innovative, and hopeful solutions. Most importantly, it will show how anyone can make a difference. Everyone can be part of the solution regardless of gender, age, color, or creed.

This book is divided into three parts. In Part 1, Claire Strom's history of hunger relief will provide a rich background. She will cover a wide range of time, culture, and geography from before the common era to the 1970s. She will discuss the reasons for hunger relief, such as famine, crop failures, war, and economic upheaval, as well as some of the solutions used to address them. Her narrative provides a fascinating journey on the role of charity, religion, and government, as well as the various attitudes toward the poor and how they were treated. While hunger relief continues to evolve today, it has deep roots in history.

The remainder of the book will then transition to my research and viewpoint during my time as a food banker. Part 2 will focus on US domestic hunger relief, starting with the formation of the first food bank in the late 1960s. It will provide insight into the formation of the Feeding America network and its evolution over forty-some years. It will offer additional insight into why food is essential beyond just filling the belly. You'll find out the facts about who is hungry through personal stories that will break widely held stereotypes and misconceptions. You will discover the many partners in the fight against hunger, their unique roles, and how they are addressing the problem.

Part 3 will describe how people are being helped: through innovation; policy and advocacy; and leadership, culture, and vision. You will also discover how all of us can be part of the solution.

My background is in building and leading food banks on a significant scale. I have experienced the nonprofit world from various perspectives and through many lenses. I started years ago as a volunteer board member of a local organization. Since then, I have served in statewide, national, and international organizations for a total of forty-three years. I have coached small nonprofits, and I have chaired boards, task forces, and numerous national committees. I have also experienced nonprofits in the role of CEO. I have advised philanthropists on their decisions to donate resources and have testified to Congress on behalf of nutrition programs and people facing hunger. I will call on my experience as the source of many real-life examples in the following chapters. I hope that almost all nonprofits can benefit from the many transferable principles I will share from my experience.

This is not a textbook, history book, or nonprofit management manual. It is a compilation of topics that have not existed before in one space and which are hopefully of value and interest to you. I realize that the food bank members across the country are diverse; there is no "one-size-fits-all" program or style of operation. They are unique reflections of their community and leadership. Beyond their missions and notable people, perhaps the only other common element is the ubiquitous banana box, the container used for millions of family food boxes. All food bank members share certain commonalities, goals, and aspirations, and together, they hope to find a permanent solution to the problem of hunger.

EMPTY PLATES

Part I

A History of Hunger Relief
by
Claire Strom

Chapter 1
Combatting Hunger

Hunger—the lack of adequate food—is ubiquitous in human history. No society has existed without hungry people and, until the last several hundred years, few have existed without periods of widespread hunger or famine. Methods of dealing with hunger are equally old and have included individual actions to increase the quantity of food, help from the wider community, religious charity, and government intervention. The fight against hunger, however, has changed over time. Urbanization increased human density, making periods of hunger dangerous to social stability and influencing governmental policies around food relief. Technological advances have improved the global food supply and made it easier to preserve and transport it. Corporations and nongovernmental organizations (NGOs) have grown and assumed an outsized role in fighting hunger. And, finally, attitudes toward hunger—and the poverty that usually causes it—have fluctuated between a belief that humans can and should eradicate need, and a sense that poverty is an inherent representation of idleness or inadequacy that should not be rewarded.

Famine in the Ancient World
In the ancient world, famine was common due to localized crop failures, and hunger was even more common. Despite the relative paucity of source materials, many of the main themes of hunger relief emerge from the accounts that remain. Individual charity was already important, and the earliest accounts of food being distributed to the hungry were

recorded for posterity on Egyptian stelae. Rich donors who had aided the population during hard times wanted to ensure that their deeds were remembered and had self-promoting messages etched in the stone, such as this one from 1700 BCE: "I gave bread to the hungry, sandals to the barefoot; I gave corn to the entire country, I saved my city from starvation. Nobody did what I have done."[1]

More common, although less easily identified across time, were efforts by the hungry to address food deprivation. In classical Greece, farmers practiced crop diversification to ensure sufficient food if one or more crops failed, and they also relied on extended familial networks to redistribute any excess food in times of hardship.[2] Such tactics were likely common throughout the ancient world and thus rarely made it into the historical record. However, more drastic and personal ways of dealing with famine are well recorded. Stories of men selling or abandoning their children and wives to reduce pressure on their food supply appear in texts from the Chronicle of Joshua the Stylite to the Book of Nehemiah. And, when food was still insufficient, they sold themselves into slavery. In Genesis 47:19, starving Egyptians asked the Pharaoh's agent, Joseph, "Why should we die before your eyes, both we and our land? Buy us and our land for bread, and we and our land will be servants unto Pharaoh."[3]

Even in these early years, governments tried to help their people during famines, largely to prevent the crime and civil unrest that inevitably accompanies mass starvation. One tactic seen after an Athenian famine around 594 BCE was the governmental cancellation of all famine-incurred debts, which meant freeing Athenians who had sold themselves into slavery to eat.[4] Other government aid focused on ensuring that available food was evenly distributed. So, from 676–678 CE, the authorities in the Roman regional capital of Thessaloniki ordered the homes of those suspected of hoarding food searched.[5]

---

1   Cormac Ó Gráda, *Famine: A Short History* (Princeton University Press, 2009), 197.

2   Ó Gráda, *Famine*, 70.

3   Ó Gráda, *Famine*, 8, 57.

4   Ó Gráda, *Famine*, 80.

5   Ó Gráda, *Famine*, 135.

Ancient governments also tried to store excess food to distribute during hard times, again with the clear motive of preventing civil unrest and ensuring their continued rule. Perhaps nowhere was this policy more developed than in Rome. Gaius Gracchus, a politician of the Republic, ensured his popularity by purchasing excess wheat and distributing it free to the citizens of Rome. Anyone was eligible, although they had to be willing to wait in long lines. This policy continued under the emperors, with periodic attempts to reduce the two hundred thousand or so people eligible for aid through some form of means test. Overall, the grain distribution was so successful at maintaining political popularity in Rome that pork, salt, and olive oil were sometimes distributed in addition, and the practice spread to the eastern capital of Constantinople. Of course, providing free food was not, in fact, free. The government paid for it with taxes on the provinces, which contributed to the provincial sense of alienation and the ultimate collapse of the empire.[6]

Religious Teachings Regarding Food Charity

Most of these early methods of preventing hunger were developed in the face of need and lacked a theoretical basis. Conceptual frameworks for addressing hunger emerged extra-nationally with the development and spread of the world's religions. All major religions address the problem of hunger and the need for charity, but the motivations behind the charity and attitudes toward the poor vary greatly. In Hinduism, the world's oldest religion, charity, like everything else, is controlled by dharma. Dharma defines everyone's responsibilities by caste, sex, age, and family.[7] Members of some Hindu castes (Kshatriyas and Vaishyas) must give gifts to fulfill their dharma, while Brahmins must receive gifts.[8] Many

---

6   Henry Hazlitt, *The Conquest of Poverty* (Foundation for Economic Education, 1996), 67–68.

7   Mark Jurgensmeyer and Darrin McMahon, "Hindu Philosophy and Civil Society," in *Philanthropy in the World's Traditions*, ed. Warren Ilchman, Stanley Katz, and Edward Queen II (Indiana University Press, 1998), 264.

8   Leona Anderson, "Contextualizing Philanthropy in South Asia: A Textual Analysis of Sanskrit Sources," in *Philanthropy in the World's Traditions*, 64.

Hindu texts discuss *dāna*, which cannot be easily translated. It refers to charity and giving and can also mean the gift itself. It also ties giving closely to dharma.[9] Gifts can come in many forms, from land to money, but food is highly valued. Indeed, the Baudhāyana Dharmasūtra says, "Food is the highest offering." Many texts indicate that failure to offer food to guests will result in the loss of spiritual merit, while also stressing the importance of giving to the needy.[10] The concept of dharma complicates Hindu charity. Those whose caste and position in life require them to give must do so selflessly or lose spiritual ground, irrespective of the recipient's worthiness.[11]

Buddhism places even more emphasis on *dāna* than Hinduism. The religion stresses poverty as a spiritual advantage, viewing a lack of attachment to worldly goods as imperative to enlightenment. Buddhist monks and nuns survive by begging; thus, practicing *dāna* demonstrates a concern for the well-being of all, as well as a disinterest in material possessions.[12] Buddhism divides *dāna* into three types: the gift of material things, including food relief; the gift of safety; and the gift of dharma, meaning education.[13] The centrality of the gift of material relief in Buddhism resulted in the religious sustaining the poor from their temples.[14] However, unlike other religions, most notably Christianity, Buddhism never developed a centralized religious structure to administer its actions and properties. Consequently, Buddhist charity is extremely localized and varies greatly from country to country.[15]

---

9   Anderson, "Contextualizing Philanthropy in South Asia," 58.
10   Anderson, "Contextualizing Philanthropy in South Asia," 70, 63.
11   Anderson, "Contextualizing Philanthropy in South Asia," 74.
12   Ananda Guruge and G. D. Bond, "Generosity and Service in Theravāda Buddhism," in *Philanthropy in the World's Traditions*, 79–88.
13   Guruge and Bond, "Generosity and Service in Theravāda Buddhism," 92–93.
14   Amara Pongsapich, "Politics of Civil Society," *Southeast Asian Affairs* (1999): 325; Tadashi Yamamoto, "The Nonprofit Sector in Japan: Historical Evolution and Future Challenges," in *The Nonprofit Sector in Japan*, ed. Tadashi Yamamoto (Manchester University Press, 1998), 89; Yoshinori Yamaoka, "The Historical Background of Japan's Nonprofit Sector," in *The Nonprofit Sector in Japan*, 19–20.
15   Guruge and Bond, "Generosity and Service in Theravāda Buddhism," 93.

Unlike Hinduism and Buddhism, Jewish charity usually took the form of money. The Talmud prescribes how charitable funds should be collected and outlines the creation of tax collectors and administrators, called gabbaim. The Talmud also specifies who should receive help and what that help should consist of, making no distinction between worthy and unworthy poor but saying that "each and every one should be supplied with what he needs."[16] The Jewish people's status as outsiders in medieval Europe increased their responsibility to the Jewish poor, as they could not rely on churches or guilds for help as the Christian population could.[17] Jewish communities in Europe, however, were also concerned that crimes committed by their poor might result in general persecution, and so, by the eighteenth century, Jewish communities had regularized poor relief, creating workhouses for the local poor, which were often funded by a poor tax. The communities expelled the nonlocal poor to try to reduce crime that might generate Christian retribution, and to conserve resources.[18]

The early Christian Church viewed poverty as inevitable.[19] God created the poor along with the rest of the world. They were God's creatures and deserved charity, a word stemming from the Latin *caritas,* which was used in the Vulgate Bible to mean Christian love.[20] Indeed, unlike some other religions, Christianity expresses a degree of suspicion toward wealth, with a strong narrative of the holy beggar and a condemnation of, if not money, then the means of acquiring it: "It is easier for a camel to pass through the eye of a needle than for a rich man to enter into the kingdom of Heaven."[21] Consequently, poor

16   Walter Trattner, *From Poor Law to Welfare State* (The Free Press, 1999), 3.
17   Derek Penslar, "The Origins of Modern Jewish Philanthropy," in *Philanthropy in the World's Traditions,* 198.
18   Penslar, "The Origins of Modern Jewish Philanthropy," 199–203.
19   Paul Fideler, *Social Welfare in Pre-industrial England* (Palgrave McMillan, 2006), 9.
20   Marjorie Keniston McIntosh, *Poor Relief in England, 1350–1600* (Cambridge University Press, 2012), 6.
21   Fideler, *Social Welfare in Pre-industrial England,* 14; McIntosh, *Poor Relief in England,* 15; Matthew 19:24.

relief is embedded in Christianity, and feeding the hungry is one of the seven corporeal acts of mercy.[22]

In practical terms, the Christian Church aided the poor in two ways. First, lay orders constructed monasteries that became relief centers by the sixth century. While providing a place where potential beggars, such as widows, could live and receive food, these orders also used donations to build almshouses and hospitals where the poor—usually the old and sick—could live for free, often being allocated food, clothing, and fuel.[23] Second, parishes acted as administrative units for poverty relief. Parish priests controlled the tithe, which had originally consisted of grain stored in a public barn where the poor could monitor its distribution.[24] Parishes also collected and distributed alms (usually in the form of money) either as outdoor relief—help to people in their own homes—or by constructing parish almshouses.[25] Similarly, in the Eastern Orthodox Church, the thirteenth-century Serbian Code of St. Sava required that the Church establish homes where the elderly, sick, insane, and orphaned could be housed and fed.[26]

Despite the early Christian Church's protestations that God created all the poor, early Christian theologians, such as St. Augustine and St. Ambrose, distinguished between the worthy poor, who could not help themselves, and the able-bodied poor.[27] The importance of this distinction was heightened with the Protestant Reformation. The Protestant denial of salvation through good works reduced the spiritual necessity of almsgiving.[28] Martin Luther disapproved of begging, which led to the Protestant stress on labor and

---

22   Fideler, *Social Welfare in Pre-industrial England*, 14; McIntosh, *Poor Relief in England*, 15.

23   Philip Gavitt, *Gender, Honor, and Charity in Late Renaissance Florence* (Cambridge University Press, 2011), 13; McIntosh, *Poor Relief in England*, 7.

24   Ó Gráda, *Famine*, 202.

25   Fideler, *Social Welfare in Pre-industrial England*, 19.

26   Miroslav Ružica, "Orthodox Christianity, the Nation-State, and Philanthropy: Focus on the Serbian Orthodox Church," in *Philanthropy in the World's Traditions*, 242.

27   Fideler, *Social Welfare in Pre-industrial England*, 16.

28   McIntosh, *Poor Relief in England*, 20.

a consequent condemnation of the unworthy poor.[29] More importantly, Luther, as well as John Calvin, revised the Catholic sacrament of good works. Instead of regarding "good works" as spiritual exercises, the early Protestants redefined them as working hard at one's profession. Working hard often led to material rewards.

Consequently, Protestants began to view wealth as a sign of election—being chosen by God. Conversely, poverty was increasingly regarded as a sign of spiritual weakness and moral failure. As one Englishman expressed in his will, it was thought that the practice of begging would lead to "whoredoms, robberies, and all other profaneness, to the great offense of almighty God who hath expressly said that there shall be no beggars in the land."[30] This shift in Christian thinking had an oversized impact on attitudes toward poverty in the Western world from the sixteenth century on.

Like in other religions, charity is central to Islam. One of the five pillars of Islam is zakat, an obligatory alms tax calculated as a percentage of a person's entire wealth.[31] Zakat is not philanthropic in many senses because it does not represent human concern for the disadvantaged. Rather, for the donor, zakat is an attempt to become closer to God by obeying his divine command to give alms.[32] Additionally, Islam has *ṣadaqa*, or voluntary almsgiving, and the Quran includes the needy among its potential recipients.[33] Usually, this charity assumed monetary form, although the rich occasionally distributed food. Such food distribution often occurred during major religious festivals and brought substantial acclaim to the donor.[34]

---

29 Ó Gráda, *Famine*, 202; McIntosh, *Poor Relief in England*, 22.

30 McIntosh, *Poor Relief in England*, 22.

31 Gregory Kozlowski, "Religious Authority, Reform, and Philanthropy in the Contemporary Muslim World," in *Philanthropy in the World's Traditions*, 281.

32 Kozlowski, "Religious Authority, Reform, and Philanthropy," 282.

33 Yaacov Lev, *Charity, Endowments, and Charitable Institutions in Medieval Islam* (University Press of Florida, 2005), 8.

34 Lev, *Charity, Endowments, and Charitable Institutions in Medieval Islam*, 36, 42.

The other main form of Islamic charity, not mentioned in the Quran, involves waqf.[35] The rich established these foundations, often with a profit-generating enterprise created to sustain them.[36] For example, in the fifteenth century, Qaytbay established a public kitchen in Medina that he financed with the profits from a commercial property.[37] Waqf served a wide variety of purposes. Some focused on improving communal infrastructure, while others educated the poor—including through institutions of higher learning that became the basis for modern universities. Others supported mosques, while some provided medical care.[38] In terms of food relief, some waqf established ribats or lodges that offered help to specific groups, from pilgrims to widows. Many, however, were established specifically to house the poor, and thus combined shelter and food relief, much like medieval Christian hospitals and almshouses.[39]

Secular Premodern Food Relief

Along with religious aid, private individuals and organizations remained important in premodern food relief. People facing famine responded with desperation and did what they could to survive. During the famine of One Rabbit (circa 1454 CE) in Mexico, many people left the capital and went to the province of Totonacapan, which still had corn. There, they sold their children, with boys fetching five hundred ears and girls four hundred.[40] Less than one hundred years later, in the face of another famine, families in Florence, Italy, abandoned their children, whom they could no longer feed, to the

---

35   Said Amir Arjomand, "Philanthropy, the Law, and Public Policy in the Islamic World before the Modern Era," in *Philanthropy in the World's Traditions,* ed. Warren Ilchman, Stanley Katz, and Edward Queen II (Indiana University Press, 1998), 110.

36   Lev, *Charity, Endowments, and Charitable Institutions in Medieval Islam,* 53–55.

37   Lev, *Charity, Endowments, and Charitable Institutions in Medieval Islam,* 55.

38   Lev, *Charity, Endowments, and Charitable Institutions in Medieval Islam,* 53–114.

39   Lev, *Charity, Endowments, and Charitable Institutions in Medieval Islam,* 115–116.

40   Ó Gráda, *Famine,* 58.

local ecclesiastically run orphanage.[41] And, when selling or abandoning children proved insufficient, starving people sold themselves into slavery, with booms in the Dutch slave trade aligning with famines on the Indian subcontinent.[42]

Another tactic of starving people was to riot for food, and this threat often spurred elite aid. During a famine in China from 1640 to 1642 CE, a Chinese official, Ch'en Tzu-lung, was returning home from the capital where he had been for the New Year. He wrote, "Along the roadside I saw thousands of starving people forming mobs; with knives outstretched." Ch'en then described how he distributed food that he had encouraged elites to purchase the year before to quiet the situation.[43] Elsewhere, people founded associations to help members during hard times, including periods of hunger. In colonial North America, fraternal organizations emerged in the seventeenth century. The first was the Scots Charitable Society in Boston in 1657, which helped members based on national or religious affiliation.[44] Similarly, in the late eighteenth century, the Japanese founded organizations called Edo machikaisho, in which neighbors donated money to—among other things—build granaries to store rice against hard times.[45]

In precolonial Africa, systems of reciprocity defended against hunger. People developed networks for religious and economic reasons, and these networks often provided food when necessary. Many African religious beliefs centered around revering ancestors. In order to be revered, a person needed descendants; thus, some wealthier people "adopted" those in need as their "children." Economically, land was readily available, but labor was not. Men could generate labor by producing children for which they needed a wife. Throughout much of Africa, bride wealth, the payment made

---

41   Gavitt, *Gender, Honor, and Charity in Late Renaissance Florence*, 34, 36.
42   Ó Gráda, *Famine*, 58.
43   Joanna Handlin Smith, "Chinese Philanthropy as seen through a Case of Famine Relief in the 1640s," in *Philanthropy in the World's Traditions*, 138.
44   Trattner, *From Poor Law to Welfare State*, 35.
45   Yamaoka, "The Historical Background of Japan's Nonprofit Sector," 22.

to a family for losing the labor of their daughter, was often made in food—usually reproducible food, such as goats, sheep, or cattle. Wealthy men and women formed additional links when they loaned parts of their herds to others to maintain in exchange for the milk and blood produced. Ultimately, the African system of reciprocity, which included a wide variety of precolonial practices involving the direct distribution of food to people in need, fundamentally differed from Western charity. In Africa, each party received something tangible from the relationship, whereas in Western charity, the rewards to the benefactor were usually spiritual.[46]

However, individual, and informal collective attempts to ensure sufficient food were often inadequate, and governments continued to play an important role in averting hunger. Governments worldwide continued to use their wealth and power to create and maintain public granaries.[47] For example, after the famine of 1460, the sultan of Kashmir canceled debts the poor had incurred to buy food. And, over three hundred years later, the nabob of Lucknow on the Indian subcontinent employed forty thousand people on public works to prevent them from starving during the Chalisa famine.[48] The Qing dynasty in China took a more direct approach when it established a system of granaries throughout the country and used canals to move grain where it was needed. When famine threatened, as with the 1743–1744 drought, food was distributed without any form of means testing.[49]

In Western Europe, the role of government in food relief expanded exponentially from the sixteenth century on. As the urban poor population grew, it exceeded the capacity of religious charities and wealthy almsgivers and threatened political stability. An additional stressor was the Protestant

46   Steven Feierman, "Reciprocity and Assistance in Precolonial Africa," in *Philanthropy in the World's Traditions*, 4–13.

47   See, for example, Juhani Koponen, "War, Famine, and Pestilence in Late Precolonial Tanzania: A Case Study for Heightened Mortality," *The International Journal of African Historical Studies* 21, no. 4 (1988): 644–645.

48   Ó Gráda, *Famine*, 80, 213.

49   Mike Davis, *Late Victorian Holocausts: El Niño Famines and the Making of the Third World* (Verso, 2001), 281.

Reformation, during which members of the lay orders often abruptly became homeless paupers in Northern Europe. The dissolution of these orders also ended their charitable endeavors, placing more strain on the wider society.

In England, where these problems emerged early and severely, the initial impulse was to ensure that all beggars were local. During the Middle Ages, towns had assumed some responsibility for their own poor, but with the increase of transient beggars, they struggled to limit costs. One solution that was attempted in the sixteenth century involved issuing badges to the local poor that entitled them to beg within the community.[50] This was coupled with attempts to expel or punish beggars who were not local residents.[51] Some town governments also provided some poor relief that was funded through taxation. Indeed, in 1588, the city of York imposed a short-lived tax on all its wealthy citizens to provide a set daily wage for paupers who were "aged, lame, and impotent, and past their work."[52]

English parishes became increasingly important bureaucratic entities outside of cities, emerging as the fundamental unit of poor relief.[53] Churches collected alms, which they then administered to the poor. With the Church under state control by the mid-sixteenth century, parliamentary legislation and royal injunctions governed the collection and distribution of funds.[54] This shift was solidified in the two Elizabethan poor laws of 1598 and 1601, which were the basis of poor relief in England until 1834, and in the United States as well until the Great Depression of the 1930s.[55] The Elizabethan poor laws articulated poor relief as a function of government and located it firmly within the parish. Begging licenses were abolished, and overseers of the poor were appointed to collect

---

50   McIntosh, *Poor Relief in England*, 44–45.
51   McIntosh, *Poor Relief in England*, 146–147.
52   McIntosh, *Poor Relief in England*, 146.
53   McIntosh, *Poor Relief in England*, 128.
54   McIntosh, *Poor Relief in England*, 129–134.
55   McIntosh, *Poor Relief in England*, 274; Gabriel Loiacono, *How Welfare Worked in the Early United States: Five Microhistories* (Oxford University Press, 2021), 12.

and distribute aid and funds.[56] Relief rarely took the form of cash given to the pauper; rather, it was usually some form of direct aid, such as housing, food, or clothing. In colonial America, overseers might pay rent and buy food for paupers, or they might set up accounts for them at local stores.[57] Some communities paid families to take in and support paupers, even auctioning them off to the lowest bidder.[58] In general, the impact of Protestantism meant that the government and society had little sympathy for the able-bodied poor, who, in the American colonies, could be bound out for labor, punished, or imprisoned.[59]

Hunger in the Age of Industrialization and Colonization

In the late eighteenth century, major changes impacted hunger and society's responses. The squalor and poverty of industrialization increased the elite fears of revolution. Meanwhile, Enlightenment ideas about individuals determining their own fate produced a moral disdain for poverty and a laissez-faire attitude toward poor relief. Nonetheless, traditional forms of individual and informal hunger relief continued in many areas.

The French Revolution of the late eighteenth century dramatically impacted how European governments understood the problem of poverty. The revolutionary government between 1789 and 1795 explicitly made poor relief a state responsibility, stating in the 1793 *Declaration of the Rights of Man and Citizen* that "public relief is a sacred debt." Unfortunately, the extensive relief program the government devised proved too expensive and was largely abandoned by the postrevolutionary government in 1796. However, the mandate laid the philosophical groundwork for developing state welfare in the twentieth-century world.[60] Moreover, the revolution cast shock waves throughout the rest of the industrialized world. The fear of social disorder and rebellion forced countries

---

56   Trattner, *From Poor Law to Welfare State*, 11; McIntosh, *Poor Relief in England*, 274.

57   Loiacono, *How Welfare Worked in the Early United States*, 35.

58   Trattner, *From Poor Law to Welfare State*, 18.

59   Trattner, *From Poor Law to Welfare State*, 22–23.

60   Fuchs, *Gender and Poor Relief in Nineteenth-Century Europe*, 39.

to reassess how they addressed poverty. Throughout the Western world, poor relief increasingly meant institution-alization, which gave governments greater control over the masses of the poor.[61]

Institutionalization also aligned with the changing ideo-logical conceptions of poverty. By the early nineteenth century, Protestantism and the Enlightenment had shifted the notion that poverty was inevitable. Rather, poverty, at least among the able-bodied, was considered a moral failing. As the New York Society for the Prevention of Pauperism stated in 1821, "No man who is temperate, frugal, and will-ing to work need suffer or become a pauper for want of employment."[62] This attitude was upheld by the economic thought of Adam Smith and others who presumed that work was available to everyone who wanted it and that workers would receive a sufficient wage.[63] These views, in turn, influenced attitudes toward poor relief. Wealth was increasingly seen as evidence of moral superiority, so taxing the rich to help those who did not help themselves was regarded as morally wrong.[64] Consequently, poverty could and should be punished, or at least relief made so unpleasant that people would do everything in their power to avoid it. In England, this notion was embodied in the 1834 reforms to the poor laws, which abandoned parish-based aid in favor of larger poor law districts and replaced most outdoor relief with institutionalization.[65] Institutions also served industrialization by providing a captive workforce. In many workhouses across Europe in the nineteenth century, beggars and vagrants were incarcerated and compelled to labor to make manufactured goods.[66] Government officials made workhouses as unpleasant as possible to deter possible inmates. Frequently, families were split up and housed in different institutions based on sex. The workhouses also had strict rules, often requiring twelve hours or more of

61   Fuchs, *Gender and Poor Relief in Nineteenth-Century Europe*, 40.
62   Trattner, *From Poor Law to Welfare State*, 54.
63   Trattner, *From Poor Law to Welfare State*, 53–54.
64   Trattner, *From Poor Law to Welfare State*, 49–50.
65   Trattner, *From Poor Law to Welfare State*, 51–52.
66   Fuchs, *Gender and Poor Relief in Nineteenth-Century Europe*, 40, 205.

labor per day and regulating all behavior, from speaking to going to the bathroom.[67]

The desire of governments to control their populations and avert social unrest was even more acute during famines. Nonetheless, governments generally still tried to avoid providing substantial direct gifts of food. Instead, they preferred implementing work projects that would not tempt the poor to excessive idleness. Thus, during the Irish potato famine of the mid-nineteenth century, beggars were put to work on infrastructure improvements, although making "malnourished and poorly clothed" people work outside did little to alleviate the death toll. Nearly 150 years later, the Ethiopian government instituted roadbuilding and land conservation projects paid in food to offset the famine facing its people. Here, a preference for male labor hurt the female-headed households, which suffered the most from the famine.[68]

By the late nineteenth century, fears of popular unrest caused by hunger had faded somewhat, and ideas about poverty had shifted again. Influenced partly by scientific discoveries about evolution, poverty was increasingly thought of as an illness in the social fabric, and the poor had to be prevented from infecting others.[69] The poor, according to social Darwinists, represented those less fit for survival, and they needed to be contained and their impact on society minimized. This thinking upheld institutionalization and laid the groundwork for the eugenics movement of the early twentieth century.

These new European attitudes toward poverty had significant global ripple effects with the large-scale colonization efforts of the late nineteenth and early twentieth centuries. During the 1876–1879 famine in India, the English viceroy, Lord Lytton, embraced a laissez-faire approach to relief, clearly stemming from social Darwinism. The official report of the famine stated that "the doctrine that in times of famine, the poor are entitled to relief . . . would probably lead to the

---

67   Fuchs, *Gender and Poor Relief in Nineteenth-Century Europe*, 203–205.
68   Thomas Keneally, *Three Famines: Starvation and Politics* (Public Affairs, 2011), 226–227; Ó Gráda, *Famine*, 214–215.
69   Fuchs, *Gender and Poor Relief in Nineteenth-Century Europe*, 13.

doctrine that they are entitled to such relief at all times."[70] The government made relief hard to obtain, with much of it in the form of public works. The men lucky enough to be chosen for work relief received a pound of rice per day. This was half the amount allocated to Indian felons, and it lacked nutritional benefit. Indeed, as author Mike Davis has noted, the response to this famine was "the personification of free market economics as a mask for colonial genocide."[71] Similar stories unfolded in Africa, where colonists allowed the devastation of droughts and rinderpest (a devastating and widespread cattle disease at the end of the nineteenth century) to run their course, propelling thousands of Africans into starvation. In desperation, these indigenous peoples left their ancestral homelands to labor for White farmers or in White mines, boosting the colonial economies while devastating traditional ways of life.[72]

Of course, individuals continued to do their best to survive hunger, utilizing a wide variety of strategies. In rural areas of Europe during the nineteenth century, people still depended on kinship groups in times of hardship. Often, extended families lived together, which offered some protection to the vulnerable.[73] When family was insufficient, other tactics were used, such as migration, which became more feasible with improvements in transportation. During the Irish potato famine, as a consequence of starvation and eviction, millions emigrated to the United Kingdom, the United States, and Australia.[74] In China during a famine in the 1870s, peasants flooded southern cities in search of food in such large numbers that the Qing government used its military to stop the flow.[75] In some regions, the Chinese peasantry organized, forcibly opening public granaries and distributing the supplies; again the government's response was to turn its army on its citizens.[76]

---

70   Davis, *Late Victorian Holocausts, 33.*
71   Davis, *Late Victorian Holocausts,* 37–38.
72   Davis, *Late Victorian Holocausts,* 100–101.
73   Fuchs, *Gender and Poor Relief in Nineteenth-Century Europe,* 80–81.
74   Keneally, *Three Famines,* 163–165, 174–177.
75   Davis, *Late Victorian Holocausts,* 70.
76   Davis, *Late Victorian Holocausts,* 71.

The nineteenth century also saw a continuation of private organizations that aided the hungry. In much of the Western world, prevalent conceptions of poverty as a social evil meant that benevolent societies focused more on reforming the behaviors seen to be causing poverty, such as drunkenness, rather than feeding the poor.[77] Improvements in communications, however, expanded global awareness of famines, meaning that private organizations could, for the first time, hear about famine in faraway places and send aid. Ironically, many charities contributed to relieving foreign hunger while refusing to encourage indolence by alleviating domestic hunger. Continuing a centuries-old tradition, many of these groups were religiously based, like the Quakers, who organized food relief to help those suffering during the Irish potato famine.[78] When famine struck India again at the turn of the twentieth century, British attention was focused on the Boer War in South Africa, and substantial aid therefore came from private groups in the United States. Kansas farmers sent two hundred thousand bags of grain from Topeka, while Black church groups and Native Americans also provided assistance.[79]

---

77   Trattner, *From Poor Law to Welfare State*, 70.
78   Keneally, *Three Famines*, 239.
79   Davis, *Late Victorian Holocausts*, 164–165.

Chapter 2
Modern Welfare and Global Efforts

The early twentieth century witnessed a fundamental shift in aid, with some governments embracing the novel idea of trying to prevent need in the first place. This phenomenon, which initially existed exclusively in Europe, was the genesis of the welfare state. Starting with Germany in the 1880s (which sought to build worker loyalty toward the new German state), countries in Europe and Latin America adopted health insurance, accident insurance, and pension schemes.[1] These were designed to reduce the chances of workers becoming poor due to accidents, illness, or old age—traditionally, the most common causes of poverty.[2] By 1901, most governments in Western Europe ran at least one of these forms of insurance, and by the end of World War I, the majority had all three.[3] In 1911, the United Kingdom introduced mandatory unemployment insurance, with other countries following in the interwar years.[4] Some Commonwealth countries, notably New Zealand and Australia, passed similar legislation.[5] A radical concept intrinsic to these government programs was the replacement of social welfare with social justice—the idea that everyone

---

1   Francis G. Castles et al., *The Oxford Handbook of the Welfare State* (Oxford University Press, 2010), 71–74.
2   Castles et al., *The Oxford Handbook of the Welfare State*, 64–65; Béla Tomka, *A Social History of Twentieth Century Europe* (Routledge, 2013), 156.
3   Tomka, *A Social History of Twentieth-Century Europe*, 156.
4   Tomka, *A Social History of Twentieth-Century Europe*, 157.
5   Castles et al., *The Oxford Handbook of the Welfare State*, 68.

was entitled to a basic standard of living and that one of the roles of government was to ensure that standard.[6]

Of course, at least theoretically, the fullest articulation of social equality was to be found in communist governments. With the Russian Revolution in 1917, Russian communists tried to implement a society of complete equality. Initially, at least, that equality meant equally bad. Russia's involvement in World War I caused the mobilization of millions, leaving many families without resources or a breadwinner. Additionally, the Tsarist government had evacuated millions of people from the combat zones of western Russia. The new revolutionary Russian government tapped traditional sources of philanthropy for assistance, but these proved inadequate. Innovative new organizations sprang up in the voluntary sector, often coordinated by rural and municipal governments. Despite their efforts, the need continued to increase.[7] The advent of the communist government ended private philanthropy and made all welfare the purview of the state. All private organizations were abolished, and their funds nationalized.[8] These political changes were undermined in 1921 by a massive famine affecting one-fifth of the Russian population and far outstripping the government's ability to deal with it. The massive death toll was only mitigated by substantial aid from the United States: governmental, through the American Relief Association led by Herbert Hoover, and religious, from the Quakers, Mennonites, and others.[9]

Thus, the early twentieth century saw a growing belief in Europe that it was a governmental responsibility to ensure a basic standard of living for all citizens. The much older concept of providing help only to the needy had not disappeared, however, and it reached a new articulation in the United States during the Great Depression. The Great Depression presented

---

6   Castles et al., *The Oxford Handbook of the Welfare State*, 78.

7   Adele Lindenmeyr, "From Repression to Revival: Philanthropy in Twentieth Century Russia," in *Philanthropy in the World's Traditions*, 314–315.

8   Lindenmeyr, "From Repression to Revival: Philanthropy in Twentieth Century Russia," 315–316.

9   Lindenmeyr, "From Repression to Revival: Philanthropy in Twentieth-Century Russia, 316–317.

an odd contradiction of starving people and surplus food, which led directly to the federal government's involvement in food aid and the creation of American welfare. World War I had sparked substantial demand for production, and American industries, such as textiles, coal, and agriculture, had expanded to meet the need. The federal government had encouraged the farming sector to increase production for the war effort, and generous loans were made available to modernize operations. After the war, shrinking demand led to falling prices, and farmers struggled to pay their debts. Their only option was to produce more, which further decreased prices. Thus, during the boom decade of the 1920s, economic hardship had already arrived in some industries.

The weakening of some economic sectors led to increased unemployment as early as 1928. This was seen as temporary and was handled through traditional municipal aid and charitable organization systems. But unemployment only worsened, increasing annually until, by 1935, fifteen million men, or one-third of the workforce, were out of work.[10] The usual sources of aid quickly became exhausted, and President Hoover did not believe in federal aid.[11] Consequently, Americans begged, ate from garbage dumps, stole from fields, looted stores, and, across the nation, died of starvation.[12]

While Hoover was still in office, solutions were necessarily piecemeal. Some were totally local. Organizations around the country, from city governments to Girl Scout troops, recognized that massive quantities of agricultural products were being destroyed because that was cheaper for farmers than shipping them to market and selling them for rock-bottom prices. These organizations found ways to pay farmers some money, and they hired unemployed workers to can food to be distributed to other unemployed workers as relief.[13] Indeed, a group of California fruit growers came up with a proto–food bank when they canned surplus fruit

---

10   Frances Fox Piven and Richard A. Cloward, *Regulating the Poor: The Functions of Public Welfare* (Vintage Books, 1993), 48–49.

11   Trattner, *From Poor Law to Welfare State*, 276.

12   Janet Poppendieck, *Breadlines Knee-Deep in Wheat: Food Assistance in the Great Depression* (Rutgers University Press, 1986), 26–32.

13   Poppendieck, *Breadlines Knee-Deep in Wheat*, 36.

and distributed it to charitable organizations nationwide.[14] State governments also took initiative in responding to the crisis. In 1931, Franklin Delano Roosevelt, governor of New York State, called a special session of the legislature to enact the Wicks Act, providing the first government unemployment insurance in the United States: a move that was copied by most states by early 1933.[15]

Nevertheless, the situation, both for the unemployed and farmers, continued to worsen throughout 1932. Farmers, unable to pay their bills, watched crops they could not afford to harvest ripen in the fields while their neighbors faced fore-closure on their mortgages. In the spring, Milo Reno started the Farm Holiday movement in Iowa, threatening to withhold crops at the end of the summer if the government did not guarantee the cost of production. The "holiday" that started in August vividly demonstrated the farmers' plight: they blockaded roads to stop nonparticipating farmers' goods from reaching the market and poured gallons of milk down the highways.[16] Poverty also brought distress to urban areas, as many city governments ran out of money and stopped paying their employees.[17] Chicago, with 50 percent unemployment, saw mass demonstrations from the Communist Party, while coal miners rioted in Kentucky.[18] Clearly, the long-established fear that hunger would lead to social unrest was playing out, and by the time Roosevelt took office as president in March 1933, the national situation was dire.

Roosevelt immediately implemented measures to help the unemployed. Most of these involved putting people to work doing a wide variety of things, including building hospitals, constructing hiking trails, staging plays, and interviewing former enslaved people.[19] Through the Federal Emergency Relief Administration, the poor could receive cash relief or food aid, although its director, Harry Hopkins, preferred work

---

14 Poppendieck, *Breadlines Knee-Deep in Wheat*, 36.
15 Trattner, *From Poor Law to Welfare State*, 279–280; Poppendieck, *Breadlines Knee-Deep in Wheat*, 57.
16 Poppendieck, *Breadlines Knee-Deep in Wheat*, 77–78.
17 Piven and Cloward, *Regulating the Poor*, 66.
18 Piven and Cloward, *Regulating the Poor*, 67.
19 Trattner, *From Poor Law to Welfare State*, 282–283.

relief.[20] All these programs took time to impact the situation, however. In the meantime, Roosevelt's strategy to solve the farming problem took effect more quickly and dramatically.

The Agricultural Adjustment Act was passed on May 12, 1933, just in time to stop another Farm Holiday strike. It established a policy that remained in effect for decades and increased unequal access to food in the United States. Basically, the federal government started buying certain commodities from farmers at a guaranteed price; it would then resell the food at a loss if necessary. This propelled the government into the food-buying and food-storing business, driving many future policy decisions. The purchase of the commodities was paid for by a tax on consumers, which was regressive and hurt the poor much more than other strata of American society. Another part of the act was to pay farmers to under-produce certain commodities to drive prices up. Theoretically, this was a strong idea, but because the legislation was passed in May, crops were already in the ground, and baby animals had been born. Consequently, in 1933, the federal government paid farmers to plow under crops and sell more than six million piglets for slaughter. The piglets, too small for the slaughter yards, often escaped and ran around Midwestern cities. There were too many to kill by traditional means, so other methods, such as electrocution, were attempted. Most problematically, the animals were too small to be processed into food and too numerous to be processed into fertilizer, so many of the bodies were burned or buried.[21] The public outcry surrounding the waste of so much food at a time of privation impelled the federal government into the business of direct food aid, with Roosevelt creating the Federal Surplus Relief Corporation in October 1933 to redistribute agricultural surpluses to the needy.[22]

A final effect of the Great Depression in the United States was the 1935 Social Security Act, which introduced pensions, unemployment insurance, and a few other safety nets, bringing the country more in line with its Western European

---

20   Poppendieck, *Breadlines Knee-Deep in Wheat*, 104–105.
21   Poppendieck, *Breadlines Knee-Deep in Wheat*, 111–114.
22   Poppendieck, *Breadlines Knee-Deep in Wheat*, 121.

counterparts.[23] However, while the United States was introducing the skeleton of a welfare state, the global trend was expansionist. During the 1930s, a disparate group of countries, including Chile, New Zealand, Hungary, and Belgium, introduced family allowances, while others expanded the safeguards they already offered.[24] In 1942, the British government published a report by William Beveridge entitled *Social Insurance and Allied Services,* which posited that the government should provide "security from cradle to grave," regardless of class or income.[25] This report proved the basis for the expansion of the United Kingdom's welfare system and the development of the federal system in Canada, and it was aspirational in other nations around the globe, although not in the United States.[26]

The years after World War II continued this universal expansion in governments' involvement in their citizens' welfare. The postwar period saw the development of the idea of universal human rights, in which food figured significantly. In President Roosevelt's 1941 State of the Union address, he listed four essential freedoms, including freedom from want, which was also front and center in the United Nations' 1948 Universal Declaration of Human Rights.[27] In Western Europe, governments added new benefits to their welfare programs, including housing support and free tertiary education.[28] Meanwhile, communist governments completely controlled the production and distribution of all goods, including food, nationalizing all welfare programs with the articulated intent of providing all with an adequate standard of living.[29]

In the United States, the presidencies of John Kennedy and Lyndon Johnson returned the focus to domestic issues, including the problem of hunger. Federal commodity distribution

23   Trattner, *From Poor Law to Welfare State,* 288–292.
24   Castles et al., *The Oxford Handbook of the Welfare State,* 71–74.
25   Castles et al., *The Oxford Handbook of the Welfare State,* 79.
26   Graham Riches, *Food Banks and the Welfare Crisis* (Ottawa: Canadian Council on Social Development, 1986), 61.
27   Castles et al., *The Oxford Handbook of the Welfare State,* 85.
28   Tomka, *A Social History of Twentieth-Century Europe,* 165.
29   Tomka, *A Social History of Twentieth-Century Europe,* 177; Castles et al., *The Oxford Handbook of the Welfare State,* 86.

had continued, but by the last year of the Eisenhower administration, families were being provided with just five foods: lard, rice, butter, flour, and cheese. Kennedy became acutely aware of the prevalence of malnutrition in some parts of the country and, in 1961, he asked the United States Department of Agriculture (USDA) to develop a food stamp program to help ensure a sufficient and healthy diet.[30] Domestic hunger continued to be a hot-button issue throughout the 1960s, with reports and documentaries exposing the plight of the poor in the United States.[31] In 1964, Congress passed the Food Stamp Act, and by 1975, food stamps were available in every county in the nation.[32]

In many countries, the 1960s represented the apogee of governmental responsibility for social welfare. However, the 1970s saw a shift in Western governments' engagement in welfare and food aid. The OPEC oil embargo of 1973 and subsequent inflation stretched governmental resources. These events ushered in fiscally conservative governments, most notably those led by Margaret Thatcher in the United Kingdom and Ronald Reagan in the United States.[33] Nevertheless, the welfare system remained intact in most of Europe even as the costs grew. Indeed, with the collapse of the Soviet Union in 1991, most Eastern European countries chose to emulate their Western counterparts, with most enacting the full-fledged European social model.[34] However, more stringent policies were enacted around unemployment benefits in the United Kingdom. Across the Atlantic, the United States, which had always been less committed to government involvement in welfare, started drastically reducing benefits, a pattern that was followed in Canada during Pierre Trudeau's administration.[35]

30 Jeffrey Berry, *Feeding Hungry People: Rulemaking in the Food Stamp Program* (Rutgers University Press, 1984), 24–25.

31 Berry, *Feeding Hungry People*, 46–48.

32 Trattner, *From Poor Law to Welfare State*, 325; Berry, *Feeding Hungry People*, 77.

33 Riches, *Food Banks and the Welfare Crisis*, 1.

34 Tomka, *A Social History of Twentieth-Century Europe*, 188–190.

35 Riches, *Food Banks and the Welfare Crisis*, 1; Beth Osborne Daponte and Shannon Bade, "How the Private Food Assistance Network Evolved: Interactions between Public and Private Responses to Hunger," *Nonprofit and Voluntary Sector Quarterly* 35, no. 4 (December

Globalization and Famine

Although the general global trend of the twentieth century was toward greater prosperity and increased government involvement in averting hunger, the period still witnessed significant famines, with an estimated cumulative death toll of around seventy million.[36] However, many of these famines differed substantially from earlier disasters, both in cause and response. With several exceptions, the worst famines of the twentieth century, including the famine associated with Mao's Great Leap Forward in the early 1960s that killed upward of thirty million people, were politically caused. Authoritarian governments—from Stalin's Russia in the 1930s to Pol Pot's Cambodia in the 1970s and Kim Il Sung's North Korea in the 1990s—adopted fiscal and economic policies that directly led to deaths from starvation.[37] Most of these regimes were also notoriously secretive, and information about the famines did not leak out to the rest of the world until significantly later.

Politics also played a role in famine relief. One of the most famous global efforts to avoid hunger and potential famine—the Marshall Plan following World War II—specifically did not apply to the communist countries of Eastern Europe. Similarly, the Reagan administration withheld aid from a starving Ethiopian population in 1984 in the hopes that famine might undermine the Marxist regime in power.[38] The United States distributed another type of aid against famine in the form of knowledge with the Green Revolution. American scientists, led by Norman Borlaug and funded by the government and private foundations, bred new varieties of staple crops engineered to increase yield. However, in Asia, the United States only made this technology and the funds to implement it available to governments willing to align themselves with the West. The Green Revolution was successful, effectively ending famine in non-communist Asia,

2006): 676.

36   Stephen Devereux, "Famine in the Twentieth Century,"
      IDS Working Paper 105, Institute of Development Studies
      (January 1, 2000), 29, https://www.ids.ac.uk/publications/
      famine-in-the-twentieth-century/.
37   Devereux, "Famine in the Twentieth Century," 22–23.
38   Devereux, "Famine in the Twentieth Century," 22.

and it helped steer those nations away from their communist neighbors and toward the West.[39] As President Nixon said in 1968, "The main purpose of American aid is not to help other nations but to help ourselves."[40]

The Soviet Union also tried to expand its influence internationally during the Cold War. It built the Aswan Dam for Egypt and agreed in 1964 to buy Cuban sugar at well above the world price after the United States refused to negotiate with the new socialist regime.[41] However, the Soviets lacked the resources to match the United States' intervention on the world stage. Since the Cold War, world powers have continued to trade food and foodstuffs for influence. For example, during the Russia–Ukraine war of the 2020s, both countries vied to provide aid to African countries, such as Malawi, to gain their backing on the international stage.[42] Indeed, with global food production and surpluses at a record high, only universal political indifference can cause famine in the twenty-first century.

In the twentieth century, food diplomacy was not the only novel way of addressing famines. Several international charities emerged toward the end of World War II to address war-caused famine. The United Nations Relief and Rehabilitation Administration (UNRRA) and Oxfam (a global NGO dedicated to eradicating poverty) joined the Red Cross to provide refugees with food and shelter. Interestingly, these NGOs were politically neutral, providing aid wherever needed. This proved important during the Cold War, when famine hit countries where geopolitical concerns prevented Western nations from helping. Thus, when the Reagan administration refused to aid the starving in Ethiopia, Bob Geldof launched Band Aid. Band Aid released a fundraising record with British musicians in 1984 and another with American artists the following year,

---

39   Andrew S. Natsios, "Foreign Aid in an Era of Great Power
        Competition," *PRISM* 8, no. 4 (2020): 104–105.
40   David Chandler, "The Road to Military Humanitarianism: How
        the Human Rights NGOs Shaped a New Humanitarian Agenda,"
        Human Rights Quarterly 23, no. 3 (2001): 687.
41   Natsios, "Foreign Aid in an Era of Great Power Competition," 104.
42   Joshua Surtees, "Growing Foothold: How Russia Donates Fertiliser
        to Deepen African Alliances," *The Guardian,* March 13, 2023.

raising more than $200 million in relief funds. The attention Band Aid brought to Ethiopia also forced governments around the globe to become more involved, resulting in $50 billion invested in addressing the famine. [43]

While politics played an important role in many twentieth-century famines, technology was also significant in relieving their impact. Agricultural technology associated with the Green Revolution helped countries like India improve their food security, hedging against future shortages.[44] Storage technology has allowed surplus food to be preserved for later use over long periods of time. Transportation technologies, such as railroads, planes, and ships, have helped move food globally to where it is most needed. Today, famine is almost exclusively confined to Africa, where the transportation infrastructure is still marginal and often compromised by violence.[45]

Charitable Partnerships and Hunger

The same technological innovations that have facilitated famine relief have also changed how countries address domestic hunger. Because of advanced food preservation techniques and large agricultural surpluses, the key problem is increasingly one of distribution—how to get excess food to needy people. Traditional charities have regained their importance, with many neoliberal Western governments reducing their involvement in social services. Additionally, the innovative creation of governmental–corporate–charitable partnerships has become prominent in feeding the poor throughout much of the world.

These partnerships emerged in the 1960s with America's national focus on hunger under John F. Kennedy and then later as part of President Johnson's Great Society. Their development was set in motion when businessman John van Hengel volunteered for a mission kitchen in Phoenix. He began gleaning fruit from abandoned orchards around town and

---

43   Chandler, "The Road to Military Humanitarianism," 681; Keneally, *Three Famines*, 269–270.

44   Devereux, "Famine in the Twentieth Century," 14.

45   Devereux, "Famine in the Twentieth Century," 13–14.

soon had more food than the mission could use, so he started distributing the excess to other agencies. He asked the pastor of the local St. Mary's Church for help, and the priest found him a building and some church funds for operating expenses. In 1967, van Hengel opened the first food bank (the St. Mary's Food Bank), relying on unused products straight from the fields and packing sheds. Quickly, van Hengel expanded his operation to work with grocery stores, asking them to donate food that could not be sold because of missing labels, dented cans, or other superficial problems.[46]

The federal government, under Richard Nixon, regarded the food bank as a good model to help address hunger. So, in 1976, they funded van Hengel's organization (now renamed Second Harvest) with $1.5 million to spread the idea nationwide. By 1984, Second Harvest had become independent of government funding and separate from St. Mary's, and it had relocated to Chicago as a nonprofit organization.[47] As the organization grew, it became more corporate, which made large food industry companies more at ease working with it, and donations grew.[48]

Despite Second Harvest's fiscal independence, government involvement in food banking continued. The recession of the early 1980s drove up the American poverty rate and coincided with the Reagan administration's drastic reduction of financial assistance through the Omnibus Budget Reconciliation Act of 1981.[49] The next year, another budget act authorized the distribution of federally owned surplus food to food kitchens and other charities.[50] Responding to hunger and once again to low farm prices, Congress created the Temporary Emergency Food Assistance Program (TEFAP) in 1983 to distribute surplus commodities purchased by the government to the needy. However, unlike during the Great Depression, many of these commodities were distributed not

---

46   Janet Poppendieck, *Sweet Charity? Emergency Food and the End of Entitlement* (Penguin Books, 1998), 112–113.

47   Poppendieck, *Sweet Charity?*, 124–125.

48   Poppendieck, *Sweet Charity?*, 125.

49   Poppendieck, *Sweet Charity?*, 82.

50   Daponte and Bade, "How the Private Assistance Network Evolved," 676.

to individuals but to charitable organizations, most notably food banks.[51] In 1988, government purchases of surplus commodities plummeted, threatening the supply to food banks. By this time, however, food banks and other charitable food providers were large enough and corporate enough that they were able to lobby effectively, and the USDA started purchasing commodities solely for distribution. This fundamentally changed the nature of the government program. Instead of purchasing surpluses to bolster the agricultural sector and adding the relatively minor cost of distributing surpluses to agencies, the entire program expenditure was now only for hunger relief. As such, it was much more closely associated with the welfare state, which the first Bush administration continued to shrink. Consequently, the program was reduced significantly during the 1990s.[52] However, it had been made permanent in 1990, with a name change to the Emergency Food Assistance Program. The funding did continue, stabilizing early in the twenty-first century.[53]

Along with providing surplus food, governments supported private food assistance in other ways. Some support was in the form of direct funding. When FoodBank Johannesburg started operations in 2009, 73 percent of its funding was from the state.[54] Globally, state and local governments often saw charities as the most effective way to fight hunger, and they therefore supported them with funds. For example, a Second Harvest survey in 1993 found that food kitchens and pantries received up to 30 percent of their income from governmental sources.[55] In Canada, some city and regional governments have donated warehouses and land to food banks.[56] A less obvious relationship between food banks and governments is based on the penal system.

---

51   Poppendieck, *Sweet Charity?*, 102–103.
52   Poppendieck, *Sweet Charity?*, 149–150.
53   Daponte and Bade, "How the Private Food Assistance Network Evolved," 679–680.
54   Daniel Novik Warshawsky, "FoodBank Johannesburg, State, and Civil Society Organizations in Post-Apartheid Johannesburg," *Journal of Southern African Studies* 37, no. 4 (2011): 817.
55   Poppendieck, *Sweet Charity?*, 122.
56   Riches, *Food Banks, and the Welfare Crisis*, 21.

Food banks and other charitable organizations often provide a place for offenders to serve their community service sentences.[57] Additionally, prisoners are used to glean fields and, sometimes, to maintain community gardens in prisons to benefit food banks.[58]

The other main partner in food banking is the corporate world. Food banks continue to rely on regular donations from grocery stores, food manufacturers, and food processors. For example, the food industry has benefited from food banks in the United States. By donating unsellable food, corporations avoid paying dump fees and often receive a federal or state tax credit.[59] Additionally, since the passage of Bill Emerson Good Samaritan legislation under Bill Clinton in 1996, corporations are legally protected from liability if any donated food causes illness or injury.[60] Corporations also benefit from the good publicity involved in donating food and in the boost it gives to employee morale.[61] More problematically, from the charitable perspective, the food industry views their donations to food banks as an opportunity to identify inefficiency in their systems. As they fix these, food sources inevitably evaporate, meaning that food bank workers must find new sources.[62] So far, they have been successful, with more food available through food banks today than ever before.

In the 1980s, another form of food relief emerged in the United States that, as with food banks, reflected the reality of hunger in a land of plenty: food rescue. These efforts were known as Prepared and Perishable Food Rescue programs (PPFRP). Helen Palit started the first one in New York City in 1982, soliciting leftover prepared food from restaurants, hotels, and conventions.[63] Some local Second Harvest food banks set up sister programs, but the organization did not adopt food rescue because of safety issues. What really propelled the growth of food rescue was the adoption of the cause

---

57   Riches, *Food Banks, and the Welfare Crisis*, 29.
58   Poppendieck, *Sweet Charity?*, 152–155.
59   Poppendieck, *Sweet Charity?*, 162.
60   Poppendieck, *Sweet Charity?*, 130.
61   Poppendieck, *Sweet Charity?*, 162-166.
62   Poppendieck, *Sweet Charity?*, 166.
63   Poppendieck, *Sweet Charity?*, 127.

by the UPS Foundation (the charitable arm of the United Parcel Service) in 1988. Through financial support and technical assistance, PPFRPs proliferated under an umbrella organization known as FoodChain. Second Harvest and FoodChain worked collaboratively on a number of initiatives, including lobbying for the Good Samaritan legislation of 1996, before merging in 1999.[64]

Although they started in the United States, food banks spread to other similar countries, especially after the recession of the 1980s resulted in governmental cuts to welfare systems.[65] Affluent countries with a sizable urban population and robust agricultural production found the government–corporate–charity model appealing. Food banks appeared in Canada in 1981, in New Zealand in 1980, and in Australia in the 1990s.[66] In Europe, the first food bank emerged in Paris in 1984.[67] Nine years later, Berliner Tafel, the first German food bank, was founded, and by 1996, the German umbrella organization Bundersverband Deutche Tafel e. V. served eight hundred different groups.[68] In the early twenty-first century, food banks expanded outside of the Western world, and the Global Food Banking Network, founded in 2006, now includes a much wider geography, with participation from countries in Africa, Asia, Latin America, and the Middle East.[69]

Food banks continue to distribute food to pantries and kitchens. However, they have developed other ways to help

64   Poppendieck, *Sweet Charity?*, 127–130; notes from Dave Krepcho's interviews.
65   See the argument in Riches, *Food Banks and the Welfare Crisis*. Anton C. Zijderveld, *The Waning of the Welfare State: The End of Comprehensive State Succor* (Transaction, 1999), 5.
66   Riches, *Food Banks and the Welfare Crisis*, 4; Graham Riches, ed., *First World Hunger: Food Security and Welfare Politics* (Macmillan, 1997), 80, 5.
67   Pilar L. González-Torre and Jorge Coque, "How is a Food Bank Managed? Different Profiles in Spain," *Agriculture and Human Values* 33 (2016): 90.
68   K. von Normann, "Contribution of Non-Profit Organizations to Reduction of Food Poverty," in *Prosperity Waste and Waste Resources: 3rd BOKU Waste Conference 2009*, ed. Peter Lechner (facultas.wuv Universitäts, 2009), 65, 69.
69   "Meet Food Banks in the Network," The Global Food Banking Network, 2023, website https://www.food banking.org/global-reach/.

hungry people. As early as 1989, Second Harvest became involved in alleviating child hunger through a program called Kids Café. This program started as part of Second Harvest of Coastal Georgia and was spread nationally by Feeding America. These cafés provide meals and snacks to children after school.[70] Over time, Second Harvest incorporated more school-based programs to address child hunger. In 1992, in response to Hurricane Andrew, Second Harvest expanded its operations into disaster relief, working with the Red Cross to move food and staff to relief centers in South Florida.[71] More recently, food banks have tried to improve the health of their offerings by soliciting more donations from farmers and becoming more involved in gleaning and gardening. Although volunteers usually do this work, some food banks strive for greater food stability by teaching poor people to garden and farm for themselves.[72] Other food banks offer programs such as job training, nutritional information, and financial education to help the needy improve their situation long term.[73] Additionally, other ventures that link surplus food to needy people have emerged. In Austria, starting in 1999, food wholesalers established the first social supermarket in Linz. These supermarkets sold low-cost food to people in need, and they could soon be found throughout Austria. These organizations even offered a mobile supermarket that served rural areas. Some also opened cafés where the poor could purchase a cheap meal.[74]

Many of these nonprofit organizations have become so large that they have expanded their function beyond hunger relief. Second Harvest, which was renamed Feeding America in 2008, started surveying hunger in the United States in 1993. No other agency had ever collected this data, which

---

70   Notes from Dave Krepcho's interviews.

71   Notes from Dave Krepcho's interviews.

72   Domenic Vitiello et al., "From Commodity Surplus to Food Justice: Food Banks and Local Agriculture in the United States," *Agriculture and Human Values* 32 (2015): 420.

73   [13] Nancy Cotungna, "Food Banking in the 21st Century: Much More than a Canned Handout: *Journal of the Academy of Nutrition and Dietetics* 102, no. 10 (October 2002): 1388.

74   Felicitas Schneider, "The Evolution of Food Donation with Respect to Waste Prevention," *Waste Management* 33 (2013): 757.

provided unique insight into the extent of hunger and poverty in the nation. The agency followed up with five more studies, yielding valuable longitudinal data showing that the hunger problem is either worsening or failing to improve. In 2008, Feeding America also launched a grant program to help its member organizations expand their outreach. The same year, the federal food stamp program changed its name to Supplemental Nutrition Assistance Program (SNAP), and Feeding America started providing education about the program to its clients and helping them complete the necessary forms to enroll.[75]

Food banks rely on the unique three-pronged collaboration of government, corporations, and charity, and their proliferation has epitomized Western governments' retreat from social justice. Many people who work in the food bank, food pantry, and food kitchen areas feel deeply ambivalent about their work, recognizing its importance but wishing to address the root causes of poverty. As Fran Ficorelli, a food pantry director in Niagara Falls, New York, said: "We need to end the hunger, not keep finding ways to find more money so we can keep feeding people."[76] By the end of the twentieth century, the existence of food banks had shifted the focus from addressing social inequality to increasing charitable work.[77] Thus, society has returned to a preindustrial mentality that poverty and hunger are inevitable and charity is a moral obligation. More recently, food banks and promoters have widened their focus, combating poverty and social inequality while feeding hungry people.

Clearly, no approach to hunger has been completely effective. While famines are increasingly rare, hunger remains in our midst in all countries worldwide. Individuals, religions, charities, and governments have all attempted to tackle hunger. Many millions of hungry people have been fed but hunger itself remains.

---

75   Notes from Dave Krepcho's interviews.
76   Poppendieck, *Sweet Charity?*, 259.
77   Warshawsky, "FoodBank Johannesburg, State, and Civil Society Organizations in Post-Apartheid Johannesburg," 819.

Part 2

The Impact of Food

Chapter 3
Food Bank Milestones Over the Years

You must do the things you think you cannot do.
- Eleanor Roosevelt

Feeding America helps provide food for hungry people in the United States through a network of meal programs, food pantries, and shelters. Two hundred food banks serving every county in the nation are members. It is the largest charity working to end hunger in the United States. In 2022, the Feeding America network distributed enough food for 5.2 billion meals.[1] The national organization has undergone a few name changes, from "Second Harvest" to "America's Second Harvest—The Nation's Food Bank Network" to its current name.

Feeding America's national office is in Chicago. It helps local members with food, funds, and capacity-building to serve more people nationwide. The national office does not physically distribute food. Instead, it secures food and grocery donations from national food companies and government agencies, which, in turn, are provided to members who handle the daily distribution. The national office also offers several other services and benefits to members, such as conducting research, raising hunger awareness nationally, managing national media

---

[1]   Feeding America Impact Report. https://www.feedingamerica. org/sites/default/file/2022-12/ FA22ImpactReport.pdf

relations, monitoring food banks, facilitating communication among members, providing program and technical subject matter information, maintaining government relations, conducting workshops and conferences, and acquiring food and funds from national donors.

The food banks that are members of Feeding America collect and distribute food to hunger-relief charities and community organizations like food pantries, soup kitchens, and shelters. In addition, most food banks distribute food through institutional partners such as schools and in partnership with the healthcare industry. They receive food that would otherwise be wasted, from donors such as restaurants and grocery stores. They operate commercial-size distribution centers. Although all food banks share a need for donors and volunteers, in other ways, they are very diverse. Some are small and located in rural areas. Others are huge and distribute tens of millions of pounds of food yearly.

To be a Feeding America member, an organization must meet several criteria. Members are under contract with the national office and serve a defined geographical area to ensure no duplicative efforts. Although under contract, each member is a separate nonprofit that governs and operates its organization. Members procure local food and financial donations and have their own governing board of directors. They must be inspected by the USDA, food industry inspection services, Feeding America, and local health departments. For members who operate kitchens, there are additional inspections required. There is a high degree of accountability required of members to uphold high standards of operation. Feeding America and its members rank high among watchdog organizations such as Charity Navigator and GuideStar. Most food banks demonstrate excellent efficiency; it is not uncommon to find that for every dollar donated, up to ten meals can be provided.

Partner agencies are distribution points for food banks that provide food directly to people. They include food pantries, shelters, soup kitchens, group homes, childcare centers, and school programs. The criteria for becoming a Feeding America food bank agency are rigorous and thorough. Organizations must show proof that they are a 501c3 charity, and they are interviewed regarding where, when, what, and whom they serve. The food bank then inspects their facilities, and if they pass, they receive an in-depth orientation on food bank services, safe food handling, and various policies and procedures. They are inspected annually and must sign and uphold a detailed agreement. An individual food bank in a rural area may have a few dozen agencies, while major urban centers often have five hundred or more. There are approximately sixty thousand partner agencies in the Feeding America network.

John van Hengel is considered the father of food banking. John was born in Wisconsin in 1923; he moved to Arizona in 1962 and joined the Cursillo movement of the Catholic Church. In 1965, John started volunteering at the St. Vincent de Paul soup kitchen, and he established programs to find food and serve the hungry. Using an old milk truck, an old-fashioned flatbed truck, and a team of volunteers, he started picking citrus at private homes and delivering the fruit to inner-city missions. As the program grew, he began looking for a centralized location where agencies could pick up the product they wanted. In 1966, John approached the priest at St. Mary's Basilica to help him find a building to act as a distribution center. The church gave him a five-thousand-square-foot former bakery donated by a parishioner.

In 1967, John met a young mother who was feeding her family by visiting dining halls and finding edible food discarded by grocery stores. She told John, "There should be a place, kind of like a bank, where you can go for food, where people make donations and others can take it out." So, the concept had a name, and, in appreciation of the church's contribution, John named his distribution warehouse St. Mary's Food Bank. In the organization's first full year of operation, John and his volunteers distributed 275,000 pounds of food in the community, consisting of citrus and food donated from local grocery stores. John named the organization Second Harvest. The name has roots in the Old Testament book of Leviticus, in which villagers and farmers were instructed to leave the outer perimeter of their crops for the poor. In the ensuing decade, word of John's program resulted in the creation of food banks in other US cities, including Baltimore, Detroit, Seattle, Los Angeles, Atlanta, Cleveland, and Tulsa. In 1976, hearing of the success of St. Mary's Food Bank, the federal government offered John an unsolicited $50,000 grant to allow food bank staff to travel across the country to develop similar collection and distribution efforts. John initially refused the grant, preferring to operate without government involvement. However, when the federal government informed him they would begin creating food banks, John accepted the grant. That same year, two critical pieces of legislation were passed to increase the flow of food from companies to food banks. The first was the national Tax Reform Act, which increased the financial incentive for corporations to donate their products to food banks. The second was a California law called the Good Samaritan Act, drafted by John van Hengel and Al Brislain, which protected food donors from legal liability when providing food to food banks. This became national law in 1996, opening the doors for millions more pounds of donated food.

In 1978, the federal government doubled John's grant to $100,000, and the reach of food banks was extended to the Midwest and southern states. The Catholic Church, along with local officials, helped the expansion. In 1979, Second Harvest was formally incorporated, with the mayor of Phoenix, John Driggs, as board chair. Other board members included representatives from three food companies: John Thomas of Kraft, Barbara Knuckles of Beatrice Foods, and Dick Johnson of CPC North America, as well as food bankers Ann Miller of Baltimore, Elijah Smith of Los Angeles, Gene Gonya of Detroit, Ann Moratta of Cleveland, and Mary Ellen Heising of San Jose. Linda Takayama of the Grocery Manufacturers of America and Karen Brown of the Food Marketing Institute were added to the board shortly after. The organization's staff consisted of John van Hengel, CEO; Bob McCarty, vice president; Mike Page, food solicitor; Ken Micetic, program director; Susan Fisher, public relations; and Al Brislain, who was involved in expansion across the country.

The first ten years of the network were chaotic while food banks tried to figure out their collective identity. There was a watershed moment—a memorable three-day meeting in Lake Arrowhead, California, where John van Hengel, his assistant Mary Lou, Jack Ramsey, and nine early food bank directors met. Rodney Bivens, one of the nine, said it was incredibly exciting because they were inventing everything. They were a bunch of strong-willed entrepreneurs, each voicing their opinions. Many decisions revolved around setting national standards for operation, policies, procedures, shared maintenance fees, geographic service areas, and more. They adopted the association model, allowing each member to be independent.

In the early days, the food bank leaders' credentials were varied, but mostly, they were not people with business backgrounds. They led or belonged to spiritual communities. They were street ministers, teachers, counselors, social workers, operators of soup kitchens and homeless shelters, activists, VISTA workers, and hippies! Back then, the food bank conferences were held in interesting venues such as monasteries. The stories of those evening get-togethers are fabled: "What happens in monasteries stays in monasteries." The qualities of the people were amazing: brilliant, creative, crazy, and welcoming. They believed in sharing problems and finding solutions, dealing with complexity and ambiguity every day. Many food bankers vacationed together, and some even married. In an interview with Rodney Bivens, he posed a question packed with gratefulness: "How can we be so fortunate to do something we love with those we love?"

The leadership of the national organization has changed over the years. Each of these extraordinary people had unique gifts to bring to the table. These included business acumen, social work, planning, policy, activism, marketing, and food industry knowledge, to name a few. One common thread that ran through these leaders was a passion for the cause. An entire book could be written on their impressive backgrounds, talents, leadership styles, accomplishments, and stories. A few have passed away; the others continue to serve in extraordinary ways in other areas. The nation owes a huge thank you to these impassioned leaders.

1976–1980	John van Hengel	2000–2004	Bob Forney
1981–1984	Jack Ramsey	2005–2009	Vikki Escarra
1985–1990	Phil Worth	2010–2014	Bob Aiken
1991–1994	Sr. Christine Vladimiroff	2015–2019	Diana Aviv
1995–1999	Debby Leff	2019–present	Claire Babineaux-Fontenot

In 1981, a significant new food source became available to food banks. During the volatile time of the recession, the federal government supported dairy farmers by purchasing their surplus milk and processing it into five hundred million pounds of cheese. The government needed to figure out what to do with the cheese. "Probably the cheapest and most practical thing to do would be to dump it in the ocean," a USDA official told the *Washington Post* in 1981. Eventually, the administration freed up thirty million pounds of stockpiled cheese to feed people suffering from hunger. President Ronald Reagan created TEFAP, which has continued to provide a substantial amount of food for food banks.[2]

In 1983, John van Hengel stepped down from his role at Second Harvest to work on spreading food banking internationally. He founded International Food Banking Services, Inc., a consulting organization, the following year. This work eventually inspired the formation of the Global Food Banking Network. Food banks now exist on six continents and in nearly fifty countries with more expansion happening.

During the 1980s, food banks in the United States continued to grow and professionalize their operations as members shared information. In 1984, Second Harvest moved its headquarters to Chicago because of tremendous growth and the desire to be closer to major partners in the food

---

[2]   Erin Blackmore, "How the US Ended up with Warehouses Full of Government Cheese," The History Channel, August 25, 2023, https://www.history.com/news/government-cheese-dairy-farmers-reagan.

industry. A wave of new food banks brought the total up to approximately 180 members. More attention was focused on improving and expanding existing programs. The organization formed three regions (Eastern, Central, and Western) to facilitate knowledge and sharing among food banks. These changes increased the efficiency of food banks dramatically, resulting in ever greater amounts of food being distributed to those in need.

The increasing professionalization of food banks changed their day-to-day operations. Margaret Linnane, my predecessor in Orlando, said that in the early days, the food bank director had to perform a variety of tasks: unloading trucks, operating a forklift, driving a truck, sorting food, managing volunteers, keeping records, managing the finances, raising money, procuring food, and recruiting partner agencies. She used to keep a pair of blue jeans in the bottom drawer of her desk so that when she returned to the office after an outside meeting, she could change quickly and jump into warehouse work. She said that eventually, as the food banks became better organized, created some systems, and raised more money, it was time to remove the blue jeans from the bottom drawer.

The expansion and professionalization of food banks in the 1980s was also reflected in the development of more programs, such as the first child hunger program, Kids Café, which began in 1989. The program provides free meals and snacks to children after school in a congregate setting. It originated at the Second Harvest Food Bank of Coastal Georgia in Savannah, after two brothers were discovered looking for food one night in the kitchen of their housing project's community center. In 1993, Second Harvest launched the national Kids Café program, which provides over fifteen million meals each year.

In 1991, then-CEO of Feeding America, Sr. Christine Vladimiroff, announced that the national network would become involved in advocacy and government relations related to federally funded nutrition programs. This initiative was met with dismay; most members pushed back, stating that their role was purely operational—getting the food in and out to people facing hunger. However, Vladimiroff persisted, and today members are all involved to some degree in advocacy. Feeding America has a second office in Washington, DC, dedicated to advocacy and policy. State associations of food banks formed during the 1990s. Some of the early ones were in California, Ohio, and Georgia.

The 1990s were transformative years for food banks, as well as for me. I originally worked in the advertising agency business and then volunteered on Miami's Daily Bread Food Bank board of directors in the late 1980s. In 1992, Hurricane Andrew hit South Florida, with devastating effects. It brought widespread destruction and loss of life and resulted in tens of thousands becoming homeless overnight. There was an unprecedented need for shelter, food, and water that lasted for months. Feeding America, its members, food and financial donors, and volunteers stepped in to help South Florida's local efforts. Support came from around the world. A couple of weeks after the hurricane, the food bank's executive director left. The board asked me to go in for a month or two as an interim until a permanent replacement was found. Little did I know how my life was about to change. I unknowingly stepped into this space and am forever thankful and blessed. Bizarrely, when I became the CEO of the food bank in Orlando decades later, I was just thirty days into the role when Central Florida was hit with four hurricanes in quick succession.

One year later, once relief efforts for Hurricane Andrew had subsided a bit, Feeding America convened its first-ever disaster relief conference. Members involved in the relief effort and national staff gathered in Chicago to document lessons learned, best practices, ideas for planning for future disasters, and so much more. This meeting created the first Feeding America disaster relief plan and produced templates for individual members. I had the honor of chairing the task force for this incredibly complex plan since I was at the heart of the storm. Other disasters followed, including more hurricanes, floods, fires, and tornadoes. Each disaster, each year, much is learned and implemented in relief and response efforts. Local, state, and national partners are now working together to help victims of disasters. Feeding America is now a significant player in relief and recovery.

The 1990s saw Feeding America and food banking discovering new ways to reach hungry people. In 1994, the School
Backpack concept was developed at the Arkansas Rice Depot
in Little Rock after a school nurse noticed that hungry students were becoming ill. Teachers and nurses were troubled
by the number of children arriving at school on Mondays
complaining of stomachaches and headaches. These students struggled to focus on their assignments or control their
behavior. The horrible discovery was that many were not
eating on the weekends, or at all, due to their household's food
shortage. Food banks created partnerships with public school
systems across America. They started to provide food packs
to the schools late in the school week so teachers could dis-
creetly distribute them to the children needing help. Teacher
feedback continues to indicate that they are grateful for the
support because they have seen positive changes. It formally
became a national program in 1996 and spread quickly across
the country, with many organizations outside the Feeding
America network adopting the model. Most food banks run
the program and provide over thirty million meals annually.

Another milestone during the 1990s was the creation of
the "mobile pantry." John Arnold, then the executive director of Feeding America West Michigan, pioneered the food
bank's *Mobile Food Pantry program*, a model that has spread
across the country as one of the best ways to bring more
food—especially healthy, perishable food—to underserved
communities. Utilized most often in areas without other
access to food, mobile pantries visit key community hubs,
such as schools and churches. They distribute food, including
fresh fruits and vegetables and personal care items.[3] John also
pioneered other creative programs throughout his career,
such as Client Choice Pantries, which allow people to choose
what they need versus receiving a prepackaged box of food.
Al Brislain, executive director of the Spokane Food Bank
years ago, commented on the value of client choice from a
lesson he learned when his area had a dramatic influx of

---

3   "Mobile Food Pantries," Feeding America, 2023. https://www.
feedingamerica.org/our-work/hunger-relief-programs/
mobile-food-pantry-program.

immigrants from the former USSR. He said the food bank spent hard-earned money on peanut butter and received numerous reports that the USSR folks were throwing the jars in the trash as soon as they left the food pantry. Al said, "They had never seen or eaten peanut butter before, and it does look gross if you're unfamiliar with it." This approach to customer service has expanded into many offerings of culturally appropriate foods.

In addition to finding new ways to reach hungry people, food banks also broadened the ways they acquired food in the 1990s. One of the most significant additions was the Retail Store Donation Program in 1999. Previously, retail donations primarily came from reclamation centers that handled damaged and surplus grocery store products. These were extensive facilities where food banks had to pick up potential donations. In 1999, Food Lion asked Feeding America for help in putting together safe food-handling guidelines and operational standards for a grassroots recovery program for perishable food at their retail stores. This program shed light on a largely untapped donation opportunity if recovery logistics could be met. As a result, the food industry continued to make significant shifts in handling surplus products, with food banks picking up directly at individual grocery stores. This meant that while previously a food bank had to visit only one or two reclamation centers once or twice a month, they now had to go to multiple store locations. In some markets, such as Central Florida, food banks picked up from three or four hundred stores weekly. This was a profound shift in logistics, operations, transportation, and facilities, and an incredible expense to the food banks. However, the return on investment was hundreds of millions of additional pounds of food annually and an expansion in the variety of products available, which now included fresh and frozen foods. A few years later, realizing the vast potential, Feeding America and its local members launched a fundraising push to upgrade and purchase refrigerated vehicles.[4]

---

[4]   "Retail Support," Feeding America, 2023., https://www.feeding-florida.org/food-access/retail-support.

During the 1990s, Feeding America also became involved in research, launching several groundbreaking studies. In 1993, it conducted the first national hunger study that exposed in detail the scope and reasons for hunger. This study involved most members nationwide in a joint research survey. Thousands of partner agencies and the people they served were interviewed. The study showed that one in ten Americans had to rely on emergency food annually. This initial work prompted ongoing research, which has helped identify causes of hunger, determined priorities for programs and public policy, and yielded countless other benefits. Research continues today, reflecting further understanding and the use of technology.

In the late 1990s, the national organization also released the Red Tape Divide, a groundbreaking national study examining the barriers to food stamp access. The study found that to receive the food stamps to which people are entitled under federal law, they had to complete an application that averaged twelve pages, and which exceeded thirty pages in a couple of states. It found that the burden placed before hungry people was not required by federal law, and it deterred people from receiving much-needed assistance. As a result of the findings and the attention they received, numerous states revised and simplified their food stamp applications, making it easier for people suffering from hunger to access food.[5]

The twenty-first century has seen food bank organizations transform into sophisticated businesses, with innovation, program growth, use of technology, and more systematic thinking leading to new strategies. Food banks are a marvel of logistics and operations, the heart of making so much of their activity and impact possible. The increased volume of food distributed, new programs, mobile pantries, disaster relief and urgency of service are all supported by incredible operational teams.

In 2004, America's Second Harvest changed its name to America's Second Harvest—The Nation's Food Bank Network,

---

[5]   Halley Torres Aldeen, Marcus Fruchter, Doug O'Brien, Kimberley Prendergast, and Eleanor Thompson, "The Red Tape Divide: State-by-State Guide to Food Stamp Applications," Issue Lab, September 1, 1999, http://search.issuelab.org/resources/the-red-tape-divide-sate-by-state-guide-to-food-stamp-applications.html.

in a move meant to identify the network concept and provide higher visibility to food banks. The most recent name change to Feeding America in 2008 "communicates the positive power of food to be a catalyst in people's lives. In essence, 'feeding' has multiple meanings—providing both a fundamental need for survival and the opportunity to enrich lives."[6]

Over the years, concerns about trust and control arose between the national office and its members. In 2001, the national board recruited leadership to prioritize establishing alignment and building a more cohesive network. Part of this involved the creation of the National Affiliate Council to provide a meaningful and effective method of ensuring that the national board of directors, the national CEO, and the members maintain collaborative and productive working relationships. This council is still active as a high-functioning body with strong collaboration among national staff and food bank members. It is one of the many strengths and successes of Feeding America.

My life, too, shifted in the new century. In the early 2000s, my wife and I became empty nesters, which opened new opportunities. I left the food bank in Miami and served a few years in the Feeding America Chicago national office as VP of business development. I led a team that developed relationships with the national food industry and solicited donations. The time was invaluable for me. I brought the local food bank perspective to the national office, hopefully making a small contribution. The national office was filled with bright people excited about the mission; however, over 95 percent had never worked in or run a food bank. The national perspective gave me insight into the phenomenal food industry in this country and a deeper understanding of the significant drivers of food donation, as well as the obstacles. I also worked with members nationwide to confront their challenges and find opportunities regarding local food solicitation. I learned more broadly what people were

---

6   America's Second Harvest Changes Name to Feeding
         America, September 5, 2008, Feeding America, https://
         www.feedingamerica.org/about-us/press-room/
         americas-second-harvest-changes-name-to-feeding-america

doing nationwide beyond food procurement. I was exposed to various innovations, programs, best practices, and philosophies that stretched my thinking. Eventually, I realized that my calling was to be closely connected to a community. As good as the national experience was, I sought a different direction and was hired to lead the Second Harvest Food Bank of Central Florida in 2004. I was fortunate to find the opportunity and became deeply involved with that community.

The twenty-first century has already seen several new initiatives to increase the amount of food available to the hungry. In 2000, FoodChain and Feeding America merged. FoodChain focused on rescuing prepared and perishable foods. At the time, this was the largest nonprofit merger ever, and it enabled those fighting hunger to work as a more unified movement. Given the increasing trend toward fresh and prepared food, the merger also allowed Feeding America to expand significantly in this area. Access to food was also improved when, in 2003, the School Pantry program began at the San Francisco Food Bank. The program helps address childhood hunger, with school-based food pantries providing easily accessible food to low-income students and their families. In 2009, Feeding America incorporated the model into the Child Hunger Strategy, making it a national program. Funding from Target helped establish the program, which has served over one hundred million meals. At the other end of the age spectrum, Feeding America created the Senior Hunger Advisory Team in 2011 to advise the national office on implementing a national strategy. In 2012, standards for these programs were established, resulting in a more intentional focus on senior food security.

In addition to finding new ways to feed hungry people, food banks have begun experimenting with ways to train people for careers that will help them escape poverty. The first few decades of the century have seen an expansion of food bank commercial kitchens. These kitchens use fresh products to create complete meals and snacks for various feeding programs. Many have evolved into culinary job training and placement programs for low-income people. Further initiatives around truck driving and distribution-center training are providing opportunities for future jobs for low-income people, thus making them self-sufficient and able to put food on their table without the need for charitable assistance.

Another Feeding America initiative emerged from the Red Tape Divide because, in 2008, Feeding America members started to assist people in receiving those federal earned benefits. At food banks, people needing food assistance can receive education about the various programs and how to get signed up. For example, the SNAP outreach run by food banks offers eligibility education, prescreening for benefits, and help completing applications. Most of the two hundred Feeding America members are providing this service in some form, equating to millions of completed applications and hundreds of millions of dollars of household benefits.

Feeding America continued to research hunger, starting its Map the Meal Gap study in 2011. It measured the number of meals needed by people facing hunger that charitable and government sources were not providing. The results led to enhanced deployment of food bank resources, new programs, and an enlightened public. This study has evolved, providing important longitudinal data. The latest study in 2021 reported a meal gap of $21,466,234,000.[7]

During the mid-2000s, the Feeding America network started to emphasize proper nutrition, leading to strong collaborations with the healthcare system, including hospitals, clinics, health departments, and health insurance companies. This groundbreaking work has gained significant momentum within the network. In addition, much work is being done to address the root causes of hunger—those environmental factors that stand in the way of adequate income. Additionally, food banks have started working alongside their partner agencies to better understand people's needs and find ways to increase the dignity of service. Much of this work is being done through an initiative called Elevating Voices.

Funding all these additional programs requires additional resources. Fortunately, the early 2000s saw an increase in national hunger awareness and a corresponding increase

---

7 Map The Meal Gap. Map.feedingamerica.org 2023

in donations. After the 9/11 terrorist attack, the economy weakened overnight. The poor became poorer, and millions in the working class were left suddenly unemployed and searching for food. A national call to action was issued to the government, corporations, partners, and the public to find one million additional meals daily to feed the newly hungry Americans. The response was phenomenal. Consequently, Feeding America led the first National Hunger Awareness Day, which is now held each September.

Along with individual giving, grants and corporate donations have also grown. Since 2000, members have received approximately $1 billion in grants. For example, in 2017, Feeding America received a $1.3 million commitment from Target. This will help establish Regional Produce Cooperatives to increase the produce available for distribution to needy people, while simultaneously reducing waste. The Regional Produce Cooperatives are enabling food banks to receive a greater variety of products more consistently and at a lower cost.

The food bank movement is still relatively new and continues to make considerable strides in expanding its impact. Leaders over the decades have realized that the food bank is an incredible platform to work from. Hunger-related issues are very broad, affecting us all in one way or another. Virtually every aspect of a community can be engaged and involved. Food banks are leveraging their efficiencies, resources, and relationships in America now more than ever.

I want to share some reflections with those of you involved in the food bank movement as it progresses. These are common sentiments shared by many people who served for multiple decades. The hope is that food banks retain a clear sense of their identity. Be aware that as your organization grows, it may become an institution, and institutions tend to focus on just maintaining themselves. Don't get bogged down in a rigid culture. Don't be afraid of taking risks with each other. Open the doors and keep that curiosity and sense of urgency. Keep the culture healthy. This is not to say to return to the "good old days" because times have changed dramatically, but don't lose that sense of solidarity, of brotherhood and sisterhood. The values of respect, trust, and stewardship never get old.

In closing this chapter, I want to share a retirement speech given by Rodney Bivens in 2016, he was the president, and CEO of the Regional Food Bank of Oklahoma.[8] He started in food banking in 1979. His speech captures and articulates, in a heartfelt way, what so many feel. It includes wisdom, advice, hard-core realism, character, values, humility, and a solid call to action:

> I wish I could tell you things have improved after 35 years of doing this. Unfortunately, I have seen the needs only continue to grow. We estimate that the food we provide only serves about 40% of the actual requirements in the 53 counties we serve. And we know that current and future market conditions will only continue to increase, rather than decrease, the need among our neighbors.
>
> Some of those challenges include our ability to get enough donated food to meet the increasing demand. This becomes more challenging every year due to factors out of our control within the food industry. Greater efficiencies have resulted in increased shelf life of the food and reduced waste, reducing the amount of excess food traditionally available for donation.
>
> I have been humbled and honored to be part of this journey we call "food banking." In a true sense, it became my life determinant of personal relationships and family. It is probably one reason I am such a strong advocate of family first at the Food Bank because of what it cost me—yes, I would do it all again.
>
> I stand before you tonight as a flawed person who has made countless mistakes in his life and whose life has been touched by the pain, challenges, and courage of those we serve. Those whose face we may never see, the voice we may never hear, or name we may never know, but knowing we may have planted a seed called "hope." A seed that will someday blossom and create new seeds of "hope." You may never see

---

8   Rodney Bivens, retirement speech to the Regional Food Bank of Oklahoma, . Used with permission.

its blossoms, you may never experience its growth, or you may never know the seed bears roots, but you can and must continue planting the seeds despite all the challenges you face.

I do not stand before you tonight believing I have achieved any level of success as long as one child goes to school on Monday morning with the pains of hunger in their belly, as long as one mother raising her children has to work two or three jobs and does not eat until her children are finished or as long as one isolated senior has to choose between getting their prescription filled and food. Yes, we can and should celebrate our small successes, but we must never become so smug or arrogant to believe we will solve hunger or poverty today, tomorrow, or ten years from now. It will take a social evolution to achieve such a feat. We can make hungry lives less harsh, tragic, and desperate and provide hope.

There are two reasons why people are hungry—behavior or circumstance, and I would submit that neither is justification for not having food on your table. My three older brothers and I were raised on a small farm in Oklahoma. We knew what it meant to be "dirt poor." As poor as we were, my parents made us believe we could be and do anything we wanted. We were not allowed to feel sorry for ourselves; we learned humbly to appreciate what we had and were told, "No matter what we did in life, we should do it to the best of our ability." My dad always told me, "You are only as good as your handshake." Mom would say, "'Can't' never did anything."

Unfortunately, the American dream is dead or dying for millions of those we serve. They live in shadows, under bridges and highways, two, three, or four families to a house or apartment, work part-time for cash or minimum wage, check us out at the grocery store, mow our lawns, clean our homes, and all too often fix and serve our food when we eat out. They deserve the right to a decent living. It begins with each of us. Several years ago, I rode with Carlos, one of our drivers. After several stops and hours on the road, we were returning to the food bank when I asked him why he worked there. He said, "Where else in America can a young black man get paid a decent wage and be thanked daily for doing his job?" He has been with us for over twelve years. And Chris, who had been working for us for several months as a selector pulling boxes marked with FFK (Food for Kids), when one day he had to pull a different box full of blue backpacks. He realized it was the same backpack he got in elementary school. He had been stealing food for himself and his younger sister, but because of the backpack, he no longer had to steal; he was now helping other children like himself. We never know whose life we may touch by the seed we plant. While I would like to tell you the story about Teresa eating balloons in the bathroom because she was hungry; 89-year-old Ruth existing off of cat food and one box of oatmeal; or Alice being afraid of being given away because they had to give their dog away because they could not afford to feed him—yes, I would like to tell you that these are the familiar stories. Still, I am afraid they are becoming the norm.

We should feel their pain when we hear about families struggling to put food on their tables. Unfortunately, many families we serve feel a sense of helplessness—living from one paycheck to the next, just getting by, one crisis to another—can be overwhelming. Our heroes are sometimes working two or three jobs without benefits; our heroes are the parents who struggle from one paycheck to the next; our heroes are the seniors who worked in our factories, protected our freedoms, and gave us the New Deal—they deserve a Better Deal, our heroes are our children who against all odds and sometimes with the pain of hungry in their bellies still go to school on Monday morning. Children do not ask to live in poverty or be hungry—they are born into it. We can educate today or incarcerate tomorrow—we have a choice.

As I near my retirement, I feel some sense of relief because I know it is time for me and time for the Regional Food Bank of Oklahoma for me to leave. I also feel a sense of despair because the challenges you face today, and tomorrow, are more significant than those we met when I began in 1979. You face historic demand for your services during many cases of flat or declining resources. We are only limited by our inability to find new resources to meet this growing demand due to our lack of vision. We must raise funds from abundance, not from scarcity. As long as one Starbucks is open, if there are funds to pay for another political race, or if we can spend millions of dollars for a sports stadium, there is enough discretionary income in this country to make sure everyone has access to food—to do less is simply immoral.

In the future, I hope that food banks do not lose the innovation and entrepreneurial spirit that has served us so well historically. As organizations become more mature, there is a tendency to become more bureaucratic, to be more alike, and to measure and compare yourself to the norms or metrics that make you look good instead of striving to be great. Alikeness does not drive innovation. Food banking must maintain its spirit of change, innovation, flexibility, and willingness to take risks. Our ability to attract individuals who want to change the world and are willing to challenge, question, and dare to ask why has separated us from others. Impacting individuals, families, neighborhoods, and communities cannot always be measured empirically. We should be careful to avoid creating the same systems we are fighting. The dichotomy of disparaging differences between our top-level executives in food banking and our entry-level employees establishes the same system that does not serve this country well. The same pay disparity in the country is being replicated in our network. We cannot fight hunger and poverty in this country if we are unwilling to challenge ourselves to be an example. If we cannot be an example of lifting people out of poverty, how can we challenge others to do so? Before I left the food bank, I had a goal: to have our entry-level pay at $15.00 per hour—I failed as it set at $13.50 per hour. We all must look internally before looking externally that we are living up to the principles that will provide food bankers a sustainable wage—to do less is unacceptable.

I have realized that we will not "food bank our way" out of hunger or poverty. It will require systemic change in this country that we find it morally and politically unacceptable that one child goes to school on Monday with the pain of hunger in their belly; that a mother working two or three jobs without benefits has to depend on charity to feed her children; that seniors have to choose between getting a prescription filled and food; and that we have the highest incarceration rate of any industrial nation in the world, especially among women and minorities who are subjected to a life of poverty when released. We need to help create a social evolution where those living in poverty are viewed as our neighbors, not as our enemies or "those people." As a commentator said, "I thought it was a war on poverty—not a war on poor people." We must never forget why we are here; we must never lose the idealism that we can change the world; and we must always set an extra place at our table for those whose faces we may never see or names we may never know.

Without food, there is no hope; without hope, there is no opportunity; without opportunity, there is no success—it begins with food. You will not solve hunger today or tomorrow, but you can help restore the American dream for millions living in the shadows. Providing food can change lives; at the end of the day, that may be enough.

Chapter 4
The Power of Food

Food is a pretty good prism through which to view humanity.
- Jonathan Gold

Remember Maslow's hierarchy of needs pyramid? The most significant, fundamental needs are at the bottom, and the need for self-actualization and transcendence is at the top. One's most basic needs must be met before one becomes motivated to achieve higher-level needs. Maslow's pyramid certainly orients food as one thing we need for basic survival. Most of us could not live longer than ten days without food. In other words, food is life! If you're hungry, everything else is affected. Food is nourishment, comfort, health, and joy. It is nourishment for our bodies, souls, and hearts.

Food, as such a life-giving and sustaining necessity, has much power. This power manifests itself in many ways. On the positive side, it gives us great pleasure and connects us to others. Consequently, food has a great deal of spiritual significance in many of the world's religions. On the flip side, the lack of food has a range of devastating effects, both physical and psychological. The importance of food has even led to it being weaponized.

The philosopher Epicurus championed the pursuit of pleasure through the enjoyment of food. He believed the senses were the most reliable source of knowledge and that "the root of all good is the pleasure of the stomach."[1]

---

[1]   Athenaeus, *The Deipnosophists: Or Banquet of the Learned of Athenaeus* (Henry G. Bohn, 1854), Book XII, Chapter 67.

Following his lead, I'll start with a story about the lighter side of the power of food. One day, while visiting one of the food bank's pantry partners, I saw a little girl with her mom. The girl was excited to see a cake, most likely donated due to some aesthetic frosting mishap. She said, "Mom, for the first time, I can have a birthday party just like my friends." Who would ever want to deny a child a birthday cake? A party? A cake is not a nutritional item, but everybody deserves a treat occasionally. At the food bank, we called things like that "sometimes food." That little girl's face personified joy, clearly demonstrating the words of writer Norman Kolpas: "Food, like a loving touch or a glimpse of divine power, has that ability to comfort."[2]

The *Washington Post* wrote a fun article about a Staten Island restaurant where the menu is created and the meals made by a group of international women, mostly grandmothers, known as the "nonnas of the world." The restaurant is so popular that tables need to be reserved well in advance. Some people only come on nights when they know that Maria Gialanella, eighty-eight years old, is cooking. The owner, Joe Scaravella, says she spends the evenings hugging people. There's much love in that room. These women are bringing their cultures forward, and at the end of each evening, they receive a round of applause from the customers. Scaravella says, "It's hundreds of years of culture coming out of those fingertips. . . . It's beautiful stuff."[3] In my interview with anthropologist Bahiyyah Maroon, she stated that "traditional cuisine is passed down from generation to generation. It operates as an expression of cultural identity. Immigrants bring the food of their countries with them. Cooking traditional food is a way of preserving their culture when they move to new places."

Food, as made by the nonnas, provides love and an essential link to cultural heritage. Case in point, my wife occasionally tries to duplicate her mother's "drop fudge." Of course,

---

2   Norman Kolpas, "At Table" Is Back – Finally! October 28, 2021.
      Whereinspirationblooms.cpm/at-table-is-back-finally/
3   Sydney Page, "This Restaurant is Run by Grandmothers,"
      Washington Post, January 24, 2023.

her mom never had a recipe written down; it was always "a bit of this, a bit of that, a swirl for a minute." Shared food is also a fundamental part of many societies. John Rivers, a Central Florida restaurateur, entrepreneur, and philanthropist, said that when he was growing up, whenever someone came over to the house—it didn't matter what time it was—his mom would serve food. It was the way they greeted each other. It was an expression of family and lifestyle. To this day, he values that cultural link and has extended this welcoming gesture through his charitable food relief work.

Here's a food bank story that shows the power of food to connect people to their culture and their loved ones. Enas Gebaly, a Second Harvest Food Bank graduate of Central Florida's culinary program, is from Egypt. At the start of the pandemic, her husband lost his job, and times became tough for them and their three children. Her extended family was still in Egypt, and she missed them dearly, so she tried to bring them closer by preparing her mom's food. The smell and taste reminded her of the love her mom had put into those delicious dishes. Her specialty is cheesecake and baklava, and her dream is to start her own bakery. Enas says, "Come eat, enjoy, drink your coffee, forget any problem in your life, and come to my store and enjoy life!"

I see food as a powerful magnet; it always attracts everyone. It's a great connector, a conduit for relationships. When we "break bread with someone," it symbolizes welcome, openness, and a gesture of friendship. People are constantly gathering for breakfast, lunch, or dinner with friends and colleagues, and attending banquets and celebrations of all kinds.

One of my memories concerning food and the connections it forges happened about thirty years ago while I led the food bank in Miami. We did a hunger study and interviewed hundreds of food recipients throughout our system. We wanted to learn more about their challenges and how we might improve our services. One of my interview locations was in the inner city of Miami, at a tiny cement-block building that looked like a bomb shelter. When I entered, a group of four or five teens from the program had gathered for the interview. I got to one question which was "What would you do if meals were not provided? Would you skip a meal?" They looked at me and said, "What are you talking about? We get chips and soda here." I was shocked. They then told me that they had wandered the streets getting into trouble until someone from the program had offered them a snack of chips and soda one day. It turned out it was a peer mentor program, and it changed their lives for the better. I never looked at a bag of chips or soda in the same way again.

Dinners have always stood out as solid connectors, whether in a professional setting or at a table with the grandchildren. I want to share three dinners I experienced with the food bank to show the power of each. The first, Dining in the Dark, is done in partnership with Lighthouse Central Florida. This dinner focuses on food in an intense way. It offers an amazing gourmet meal prepared by professional chefs in total darkness. They are served by members of the Orlando Police Department's SWAT team—who navigate the pitch-black room outfitted in night-vision goggles. As you enter the dining room, you must keep your hands on the shoulders of the person in front of you, and a SWAT team member leads you. Blackout curtains and doorway tunnels create darkness where you cannot see your hand in front of your face. Guests are not told what is on the menu. It is tough to know where the wine bottles are on your table, let alone know what's white or red, and then to pour your own without spilling it. After dinner, the lights slowly appear, and diners offer to share their experience. These reflections on hunger and blindness range from humorous to deeply emotional.

Another memorable dinner experience was the Jeffersonian Dinner. I hosted one of these, focusing on the issue of senior hunger. According to the Purpose Generation website:

The Jeffersonian dinner tradition began at Thomas Jefferson's home in the 1800s. Jefferson was known to invite the thought leaders and influencers of his day to share in conversation. And there was a distinct characteristic to these talks that made them special: Jefferson and his guests engaged in one single thread of conversation, with only one person speaking at a time. By doing so, they unlocked the power of their collective wisdom. [4]

I invited an eclectic group that included a sociologist, anthropologist, film producer, architect, futurist, and ex-Disney Imagineer, to name a few. What great perspectives! I intentionally did not invite any subject matter experts on the topic of senior hunger. In the end, much enlightenment occurred for all. Indeed, a few attendees became involved with senior hunger in their own unique way. This is an excellent example of harnessing power from the community around such an important topic.

The last example of a food bank dinner I will share was hosted by Walmart as part of its ongoing commitment to hunger relief. We invited a small group of people facing hunger and a few food pantry partners so we could break bread together and learn firsthand about their struggles, ideas, hopes, and dreams. It provided the guests with a sense of dignity and value. Food was the connector. A unique feature of the dining was the beautiful wooden table provided by Walmart, which was custom made by an anti-hunger carpenter and craftsman. The food bank was the fortunate recipient of that masterpiece.

4   ThinkLab. ThinkLab's Visionary Jeffersonian Dinners Forge New Paths for Local Design. Amanda Schneider. 7/27/2023. Thinklab. design/join-in

Chapter 5
Faith and Food

Good food brings people together and nourishes the soul.
Author unknown

In part 1, Claire mentioned the tie between religion and food. Food is so vital to human existence and has such immense power that most religions have developed spiritual rituals and beliefs around it. In most religions, there's a moral command to pay attention to your neighbor's needs, primarily food. If you love your neighbor as yourself, you will provide for them. The Golden Rule says: "Do unto others as you would have them do unto you." Feeding the hungry is part of that belief, which exists in religions such as Bahá'í, Jainism, Judaism, Islam, Buddhism, Sikhism, Zoroastrianism, and Christianity.

For me, the rituals were Catholic. I remember Friday nights in the mid-1950s when my little sister and I would run around the empty dance floor as we waited for dinner at the Polish Falcons Club in Erie, Pennsylvania. We would run for the table when the French fries and fried perch platters were served, a table surrounded by sparkling red vinyl–covered chairs with chrome trim. Our parents would take us there often during Lent, a time in the Catholic Church when you were not supposed to eat meat on Fridays as a sacrifice offered to God. As a kid, I thought this was a special treat. After dinner, my sister and I always claimed we had to go to the bathroom; instead, we were taking a sneak peek into the smoky bar that reeked of stale beer and had a few ancient duckpin lanes with teenagers acting as the pinsetters. As a kid, I was mystified about how we couldn't eat meat on Fridays, but adults could smoke and drink beer. Fast forward to my first Holy Communion when we were taught that the priest was offering the body and blood of Christ. Try wrapping your head around that at seven or eight years old. Our faith traditions are deeply connected with food in many ways, no matter what religion we practice.

The intersection of faith and food is deeply connected to the hunger relief movement. The member food banks of Feeding America distribute food through a network of sixty thousand partner agencies: food pantries, shelters, soup kitchens, and various programs. Approximately 75 percent of those partners are faith based, crossing all denominations and belief systems. Their work is deeply rooted in religious traditions established centuries ago. These partners range in size and scope of services. Some smaller organizations distribute a dozen boxes of food per week, while larger centers feed hundreds of people every day. They can be small inner-city churches, or major charities like Catholic Charities and the Salvation Army. Seventh-Day Adventists, Islamic mosques, and Jewish synagogues are all involved. While each of these faiths has its beliefs, what unites them all is an agreement to feed people in need. Food is indeed a bridge. At a Just Food conference, an organizer stated, "It's a miracle that such a diverse group of people is a community, these folks who wouldn't normally seek each other out. Food draws us together, the act of sharing a meal."[1]

Food plays a central role in the beliefs and practices of the three major Abrahamic religions: Judaism, Islam, and Christianity. The Mishnah Torah, a rabbinic law code, states in 6:6, "If a stranger comes and says, 'I am hungry. Please give me food,' we are not allowed to check to see if he is honest; we must immediately give him food."[2] The Midrash Psalms state, "At the time of Judgement in the Future World everyone will be asked: 'What was your occupation?' If a person answers:

1   Just Food Inc. 2018 info@justfood.org
2   Mishnah Torah,rabbinic law code, 6:6,

'I used to feed the hungry,' they will say to that person: 'This is God's gate; you who fed the hungry, may enter.'"[3] These are just a few verses that direct followers to help the hungry. Judaism is rooted in social justice.

Dinners are typically significant in religious practices, reinforcing history, beliefs, and celebrations. During Passover at a Seder dinner, everyone is welcome at the table. The dinner and the Passover symbolize that the Jewish people must pass over from slavery to freedom, and when the world is redeemed, people will not be hungry; everyone will be fed. The different foods within the meal all carry symbolism, one of several examples being the dipping of parsley in salt water to represent the tears of the Jewish people.

Judaism also has a vital ethnic and cultural component to it with regard to food. Jews have lived in different parts of the world, but regardless of the place, they've taken that culture, borrowed from it, and given it a unique Jewish twist. Initially created in Poland, bagels were not a Jewish food, but they eventually transitioned into one. Potato pancakes were German, and the Jews made them into latkes.

When I asked Rabbi Steve Engel of Central Florida's largest Reform Jewish congregation about Judaism and food, he emphatically stated, "Food is everywhere in the Torah and Talmud." The Talmud mentions that "without flour, there is no Torah"; without it, a community cannot exist. Rabbi Engel noted the Torah did not say "wheat" but specifically "flour," which is significant. Flour is a product of human beings and God working together. God provides the wheat. The provision of food, by taking the wheat and making it into flour, occurs through a divine partnership between God and human beings.

Islam also focuses on food, especially as a form of charity. According to the hadith, the Prophet Muhammad said, "Feed the hungry, visit the sick, and set free the captives."[4] Within the Quran is the Surah At-Tawbah, which states, "Go make charity in giving as a sign of faith and success on Judgement Day in the hereafter." Imam Muhammad Musri, the president

---

3   Midrash Psalms, 118:17, trans. Danny Siegel, Sefaria, https://www.
        sefaria.org/sheets/115436?lang=bi.
4   Sahih Bukhari Vol. 7, Book 70, Hadith Number 552.

of the Islamic Society of Central Florida and the national president of American Islam, explained that within the Quran, zakat is the third pillar of Islam and means giving or charity. The poor are to be helped. When zakat is practiced faithfully, both the giver and the recipient are purified. Imam Musri points out that if your neighbors are in need, it is common to say, "I'll pray for you." What good is that? The scriptures are clear; you give to them.

Along with food as charity, Imam Musri emphasized that, in Islam, "what you eat impacts your soul. It affects how you think and how you behave, how you connect to God, and how you treat others around you. Your spirituality is very closely linked to what you eat."[5] He ended our interview with this statement: "To go to paradise: feed people, greet with peace, pray at night."

Like the other monotheistic faiths, food is important in Christianity, both in terms of charity and in terms of food traditions. The Bible contains numerous verses and stories that direct Christians to help feed the hungry. Reverend Jim Wallis tells a story of when he and a few friends in seminary performed a study on the extent to which the Christian tradition focused on taking care of people in need. They took a copy of the Bible, cut out the paragraphs related to the topic, and discovered so many holes in the Bible that it barely kept together. For example, Proverbs 22:9 says, "The generous will themselves be blessed, for they share their food with the poor." In Matthew 25:40, Jesus clarifies what separates believers from nonbelievers: "The King will reply, 'Truly I tell you, whatever you did for one of the least of these brothers and sisters of mine, you did for me.'" The followers were reminded by Jesus through His words in Matthew 25:35: "For I was hungry, and you gave me something to eat." Similarly, when I interviewed Joel Hunter, former pastor, author, and member

---

5   "Quranic Reflection No. 310 Āyat 22:51—Food and Spirituality," The Academy for Learning Islam, 2023. https://academyofislam.com/quranic-reflection-no-714-ayat-2251-food-and-spirituality/.

of the White House Office of Faith-Based and Neighborhood Partnerships and the Commission on Accountability and Policy for Religious Organizations, he referred to Leviticus 23:22, which commanded the people in the villages to leave the outer part of their crops for the poor.

In scripture, there are numerous references to Jesus feeding people. In the New Testament, so much was centered around people eating with each other. People didn't just gather in synagogues to hear the Word; they had a shared meal. Jesus often dined with people who were considered unwanted in society, and through food, relationships and change happened. One of the most iconic paintings is Leonardo da Vinci's *The Last Supper*, depicting Jesus and the twelve apostles gathered for dinner. Jesus presents himself as "the bread of life," the basis of Christian belief. He offers bread and wine to symbolize His body and blood. This offering is made daily in every Catholic church and frequently in other Christian denominations.

In Christian tradition, Easter is the holiest of celebrations, commemorating the resurrection of Jesus. Easter food traditions are steeped in symbolism. The foods vary worldwide; however, a few are enjoyed in many places. Lamb symbolizes the sacrificial lamb of the first Passover, as well as Jesus, the Lamb of God. Lambs also represent spring and new life. Bread is perhaps the most essential item on the Easter table, appearing in nearly every cuisine. Bread represents life. A traditional food in several parts of the world is the hot cross bun, a soft, spiced roll made with fruit and marked with an icing or dough cross. And don't forget the ubiquitous Easter egg, representing new life.

I recall a Catholic church my wife and I belonged to in Hollywood, Florida, back in the 1980s and its commitment to hunger relief. The church was led by Fr. Sean Mulcahy, a five-foot-high leprechaun with a huge heart for the less fortunate. He held an annual "Country Fair" with all proceeds going to feed needy people; it raised more than $100,000. The event bonded parishioners through their volunteer work at the fair; everybody had a role. Sean was very hands on in his service to the poor. He would pull up to a loading dock at the food bank with a thirty-year-old semitruck and forty-foot trailer and load it with food. What an inspiration; I don't know how he could see over the dashboard.

Food can play a significant role in communal, mystical, and spiritual practices; it has the power to nourish both body and soul. Here's a powerful example told by an elder in the Black church. "On my mother's side of the family were descendants of enslaved people from the Maryland Eastern Shore. They developed their contemplative practices around meals. I understand how people who work around crabs and bluefish can develop spiritual practices to marry full stomachs to piety. The informality of kitchen tables replaced confessionals, and important decisions were made as salmon cakes were shaped and collard greens were cut. Sunday meals were open to anyone who wanted to come. Those without families, those down on their luck, would appear for the standard fare."[6]

In practical terms, a religious focus on the importance of food and feeding the hungry has made faith communities important in broader charitable endeavors. As a food banker, I sincerely appreciate how these communities reach out, especially during disaster relief. I've worked side by side with many organizations, such as Seventh-Day Adventists, Catholic Relief Services, Mennonite Disaster Services, United Methodist Committee for Relief, Jewish Coalition for Relief, Lutheran Disaster Services, and Samaritan's Purse. This collective power has brought relief to millions of people over the years. For example, Catholic Relief Services expresses that their "vision is to promote social justice, solidarity and compassion through timely emergency and recovery actions that address the needs of the most vulnerable."[7] This aligns wonderfully with the work of food banks. When Hurricane

6   Barbara A. Holmes, "Feeding Spiritual Hunger," *Joy Unspeakable: Contemplative Practices of the Black Church* (Minneapolis: Fortress Press, 2017), https://doi.org/10.2307/j.ctt1tm7hhz.

7   "Emergency Response & Recovery," Catholic Relief Services, 2023. https://www.crs.org/our-work-overseas/program-areas/emergency-response-and-recovery.

Andrew hit South Florida in 1992, the devastation was overwhelming in scope and scale. No single organization could provide all the relief needed; dozens did so along with the government. The food bank valued the volunteer base that Catholic Relief Services brought in from outside Florida to help distribute the food that food banks provided. The organization offered their church properties throughout the area as distribution sites, as these were natural gathering places for stricken people. Daily communication about what was happening in different locations was critical, and having the link to Catholic Relief Services was vital in identifying the high-priority areas for food distribution. Their collaborative spirit has been deeply appreciated over the long haul and from one disaster to another.

Another example is working with the Mennonites; they do their work quietly and humbly, often remaining invisible to the public or the media. The Mennonite Disaster Service states that it "recruits, organizes and empowers volunteers to repair and rebuild the homes of those impacted by disasters in the United States and Canada. As we … serve people who would not otherwise have the means to recover, our goal is to restore hope and bring people home."[8] During the Hurricane Andrew relief efforts, the food bank identified a damaged shopping center near ground zero as a potential food distribution site. However, the building was uninhabitable. The Mennonites restored the building for our use, making it possible to expand our impact beyond the hurricane relief effort.

The faith communities have been and will continue to be critical partners in feeding the hungry; however, a deceptive myth claims "the churches can do it all." According to Robert Parham of Good Faith Media, "The idea that churches can tackle national poverty, take care of those who are ill, and rebuild communities after natural disasters requires a spoonful of bad moral theology and a cup of dishonesty."[9]

---

8   "Beyond the Project," Mennonite Disaster Service, 2024, , https://mds.org/.

9   Robert Parham, "'Let the Churches Do It' is a Deceptive Myth," Good Faith Media, May 24, 2011, https://goodfaithmedia.org/let-the-churches-do-it-is-a-deceptive-myth-cms-17948/.

The government spends billions of dollars on disaster relief with the incredible support of faith communities. Putting the full burden on the faith community would result in the long-term suffering of millions of people. As it is, the spiritual importance of food, together with doctrinal mandates to provide charity, has made faith groups vitally important in the fight against hunger.

Chapter 6
The Shadow Side

Out of the mountain of despair a stone of hope.
Martin Luther King Jr.

The power of food is not always positive. Food is so powerful that every day, people experience the ill effects of having too little, too much, or the wrong kind. In the United States, we have endless choices; food surrounds most of us. Yet, ironically, we have thousands of food deserts across the country, and too many people lack access to healthy foods. Those deserts are also "food swamps," flooded with fast-food restaurants and, mostly offering cheap, high-calorie, low-quality, unhealthy foods. I've heard stories about people collecting ketchup packets and adding water to them to make tomato soup to stretch their budgets. When someone is in survival mode, there is no time for all the enjoyment of food, especially when they don't know where the next meal comes from. Not only are these folks missing the joy and fun of eating well; they are also experiencing the powerful stigma attached to not having food.

Take, for example, the case of Mike Eagen. Mike, one of the past board presidents of Second Harvest Food Bank of Central Florida, is a very successful professional in the financial industry. He became involved with the food bank because of his harsh childhood memories of insufficient food. His parents were so embarrassed about lacking money to buy food that they would send Mike to the store with their food stamps to shop. Mike tells of going to school with a small paper bag filled with tissues, so it looked like he had lunch for the day. Hunger in the United States stands in stark contrast to the abundance of food many of us enjoy. To highlight this point, I recall a food bank staff member sharing a story about the person bagging his groceries at a grocery store. The food banker learned that putting food on the table was tough for this store employee. He was in the middle of a store with approximately forty thousand food items, surrounded, and was having trouble getting sufficient nutritious food for himself. The impact of lacking access to adequate food goes beyond affordability; it often carries deep negative effects of shame, embarrassment, and other psychological impacts.

In the middle of all this craziness around food are children—the most vulnerable population in food insecurity. As many as nine million children in the United States live in "food insecure" homes. How is this possible in a country as wealthy as the United States of America? Our future potential as a country is compromised.

According to the Feeding America website, a child "who doesn't get enough to eat—especially during their first three years—begins life at a severe disadvantage. Children facing hunger are more likely to be hospitalized, and they face higher risks of health conditions like anemia and asthma.... Children facing hunger may struggle in school—and beyond. They are more likely to repeat a grade in elementary school, experience developmental impairments in language and motor skills, and have more social and behavioral problems."[1] It is challenging for school teachers to get their attention; kids aren't focused or listening if they are distracted by hunger. A Spanish proverb summarizes this well: "The belly rules the mind." One study states that 60 percent of teachers regularly buy food for their students who are not getting enough to eat.[2] Food is more than just fuel for their bodies; it can provide a

1   "Facts about Child Hunger in America," Feeding America, 2023 https://www.feedingamerica.org/hunger-in-america/child-hunger-facts.
2   Lisa Stark, "Teachers Spend Hundreds of Dollars to Help Feed Students Who Are Hungry," *Education Week*, August 9, 2017.

pathway to a lifetime of possibilities for these kids. They are our future nurses, doctors, teachers, firefighters, and astronauts. Childhood food insecurity affects us all.

A child food insecurity report[3] states that hunger is a health, educational, workforce, and job readiness problem. Jack P. Shonkoff, MD, the director of the Center on the Developing Child at Harvard University, said, "The health development of all children benefits all of society by providing a solid foundation for economic productivity, responsible citizens, and strong communities." The health dimensions of food and proper nutrition are critical to a healthy lifestyle. The old saying goes, "Let food be medicine and medicine be thy food." That statement was true in the past, and it is true today. The healthcare industry is making significant strides in partnering with organizations such as food banks to focus on diet-related diseases such as hypertension, diabetes, and heart problems among the lower-income population.

Food banks are partnering with schools to help kids with nutrition through programs such as weekend food packs for elementary students, food markets at the middle and high school levels, and after-school programs for various ages. Food distributions are also held during extended holiday breaks, especially during summer. I recall a high school principal in a low-income area telling me that the food bank food had reduced the number of 911 calls the football coach had to make at practices after school. Athletes had been passing out due to a lack of nutrition.

College students also face hunger. According to the Feeding America website, "the demographic makeup of the college student population has changed in recent decades." A 2018 report from the Government Accountability Office found that most college students (71%) have one or more of the following characteristics: not receiving financial assistance from family, working part- or full-time while they are in school, and acting

3   "Child Food Insecurity: The Economic Impact on our Nation," Feeding America and The ConAgra Foods Foundation, 2009, https://www.nokidhungry.org/sites/default/files/child-economy-study.pdf.

as caretakers.[4] There are a variety of demands on the funds of both traditional and nontraditional students, such as tuition, housing, and books, and food is often low on their list. Inflation causing food prices to increase is also a factor. I spoke to a college president who mentioned that students suffer from a confidence loss without food. In response to college students' needs, food banks have explored new ways of providing access to food. These include helping students apply for SNAP and setting up mobile pantries near campuses to distribute food.[5]

Jeannie Kiriwas, the associate director of the student union at the University of Central Florida, mentioned that in the past year, seventy thousand pounds of food were provided through their pantry called Knights Helping Knights, with 350–400 students per day accessing the food. She estimates that 30–40 percent of the school's student body is food insecure. Jeannie introduced me to two students, Anne and Alberto, who have utilized the campus food pantry.[6] They come from lower-income families and are determined to succeed. Anne is completing her master's degree in health science and epidemiology. Alberto is majoring in mechanical engineering and aerospace. Neither of them wants to ask their parents for more money because they know their siblings back home need support. When they started college, neither had a car or nearby access to a grocery store, and the bus system was inadequate for their needs. Both found part-time employment, working twenty hours a week, but when COVID-19 hit, those hours were cut. With only $15 per week for food, Anne's breakfast was typically a banana and water, with ramen noodles for Alberto. He said that hunger is one of those sensations you can't push to the side and ignore, as much as you try. The students' lack of food led to a lack of energy and focus in the classroom. Eventually, they learned of other students in the same predicament, and they started pooling their meager dollars and crowding into one vehicle

4   "College Student Hunger: Statistics and Research," Feeding America, 2023, https://www.feedingamerica.org/research/college-hunger-research.
5   "College Student Hunger: Statistics and Research."
6   Names changed for anonymity.

for a trip to the store. Anne and Alberto somehow increased their work hours and spent significant time getting to and from work in shuttles. They are not out of the woods yet. Just think about what they would be capable of if they were thriving instead of focusing on basic survival.

So far, several dimensions of the power of food have been discussed: food is a basic need, a comfort, a spiritual manifestation, and a stigma, and the lack of it affects the most vulnerable children. Unfortunately, the sacredness of food can become profane. Sometimes it is used as a weapon, a despicable use of power. During wars or political disputes, food is sometimes used to punish and control other countries' welfare. A country's welfare correlates directly with the interest of its people; therefore, each country wants to have an appropriate food supply for its citizens. One recent example is Russia's war in Ukraine. The war's local impact on the food supply in Ukraine is terrible enough, but it also has a global impact. Agriculture represents 41 percent of Ukraine's exports, equivalent to $27.8 billion annually.[7] As of March 2023,, at least three growing seasons in Ukraine have been lost. Any hope of a successful harvest seems far in the future. According to a *New York Times* article, "Farmers who choose to climb into their tractors and work their land risk death or dismemberment by mines, shells, and other ordnance that litter the fields."[8] Grain that has been successfully harvested cannot be moved out of the country, and millions of tons are waiting for export.[9] World Food Program Executive Director David Beasley stated in a recent UN Security Council debate, "Failure to open those ports . . . will be a declaration of war on global food security. . . . And it will result in famine,

---

7   "Ukraine Agriculture Production and Trade," Foreign Agricultural Service, USDA, April 2022, https://www.fas.usda.gov/sites/default/files/2022-04/Ukraine-Factsheet-April2022.pdf.

8   Michael Schwirtz, Stanislav Kozliuk, and Ivor Prickett, "In Fields Sown with Bombs, Ukraine's Farmers Risk Deadly Harvest," *New York Times,* March 12, 2023.

9   Mark Green, "Forty Percent of the World Food Program's Wheat Supply Comes from Ukraine," Wilson Center, June 2022, https://www.wilsoncenter.org/blog-post/forty-percent-world-food-programs-wheat-supplies-come-ukraine.

destabilization, and mass migration worldwide."[10] The combination of rising food prices and supply disruption from the war has international ramifications and has resulted in food shortages. As Beasley also said, "In many countries, we are forced to make the heartbreaking decision to take food from hungry children to give it to starving children."[11] Thus, the victims of Russian aggression are not only found in Ukraine, but also in other areas around the globe.

Simultaneously, the humanitarian calamity in the Gaza Strip due to the Israel/Hamas war is wreaking havoc. According to the Integrated Food Security Phase Classification, 1.1 million people in Gaza – roughly half the beleaguered territory's population – are expected to face catastrophic levels of hunger and starvation. "We haven't seen the rate of death among children in almost any other conflict in the world," Catherine Russell, head of the U.N.'s children agency stated on a recent broadcast. "I've been in wards of children who are suffering from severe anemia malnutrition, the whole ward is absolutely quiet. Because the children, the babies don't even have the energy to cry."[12]

Closer to home, the United States instituted trade tariffs in 2018 on the majority of Chinese products, which continued under the Biden administration. China and other nations placed retaliatory tariffs on US goods in turn. While not anywhere nearly as devastating as the Russia–Ukraine War, this trade war offers an example of the power of food on an international scale, with the United States weaponizing food. This political action hurt American agriculture, and it also raised consumer prices for food in the United States. A USDA study found that, over an eighteen-month period, the retaliatory tariffs reduced US agricultural exports by $27

---

10   "Failure to Open Ukrainian Ports Means Declaring War on Global Food Security, WFP Chief Warns UN Security Council," World Food Programme, May 19, 2022, https://www.wfp.org/news/failing-open-ukrainian-ports-means-declaring-war-global-food-security-wfp-chief-warns-un.

11   Jennifer Rankin, "Africa Warns of Food Crisis due to Russian Blockade of Ukraine's Ports," *The Guardian*, London, May 31, 2022.

12   Israel's war on Hamas brings famine to Gaza, Ishaan Tharoor, March 19, 2024, Washington Post

billion.[13] Eventually, the United States "negotiated more than 50 trade agreements to boost market access and exports for American goods, ... provided $30 billion in support to farmers and ranchers during COVID-19," and, with the USDA, distributed "more than $4 billion for the innovative Farmers to Families Food Box program," reaching millions of people suffering from a loss of income at that time.[14] In addition, billions of dollars of food that could not be sold to some foreign markets were distributed through Feeding America and other domestic organizations. After all, there was a bit of a silver lining to this debacle.

To close out this chapter, I'll shift gears to briefly address a rapidly growing phenomenon—food sovereignty. It is yet another dimension of the power of food, seeking to provide more power to the people. A declaration at the first global forum on food sovereignty in 2007 defined it as "the right of peoples to healthy and culturally appropriate food produced through ecologically sound and sustainable methods, and their right to define their own food and agricultural systems. It puts the aspirations and needs of those who produce, distribute, and consume food at the heart of food systems and policies rather than the demands of markets and corporations."[15] While the idea of food sovereignty is novel, it reflects the age-old practices of many indigenous peoples. Communities that were able to produce sufficient food to survive could weather economic downturns more successfully than those who relied on trade or purchase to feed themselves. Recently, American Indians and Alaska Natives

---

13   Erica York, "Tracking the Economic Impact of U.S. Tariffs and Retaliatory Actions," Tax Foundation, July 7, 2023, https://tax-foundation.org/research/all/federal/tariffs-trump-trade-war/#:~:-text=The%20United%20States%20is%20currently,worth%20of%20imports%20from%20China; Alex Durante, "How Tariffs and the Trade War Hurt U. S. Agriculture," Tax Foundation, July 25, 2022, https://taxfoundation.org/blog/tariffs-trade-war-agri-culture-food-prices/#:~:text=A%20U.S.%20Department%20of%20Agriculture,percent%20and%205%20percent%20respectively.

14   "Land and Agriculture," Trump White House Archives, 2022, https://trumpwhitehouse.archives.gov/issues/land-agriculture/.

15   "Declaration of Nyeleni," Global Forum on Food Sovereignty, Mali, 2007.

have been reintroducing traditional foodways to empower themselves and protect the environment.[16] Michael Pollan stated, "Continuing to eat in a way that undermines health, soil, energy resources, and social justice cannot be sustained without eventually leading to a breakdown."[17]

I consider myself the eternal optimist; however, the idea of everyone producing their food, or even a meaningful amount of it, is unlikely. In today's society, there are just so many food options for most of us, and convenience seems to drive so many decisions. Most people are unlikely to put on their overalls and get their hands dirty. Nor do they have the time. I am hopeful for the movement; perhaps smaller communities and villages may have a better chance of wrapping their heads and hands around the challenge of achieving sovereignty with their food. Hopefully, there are initiatives on a large scale somewhere on the planet that are being watched for replication.

Nowadays, food seems to be getting much attention, whether it's about the latest restaurant opening or food truck, growing your own in the yard, or bringing food to people in need. During the past ten years, while I was in food banking and convening various groups around our mission, I was encouraged by the younger generations and their passion and enthusiasm. I remain hopeful that they will take the power of food to a new level, with even better outcomes for all.

---

16   "Why Food Sovereignty Matters," US Department of the Interior, Indian Affairs, 2023, https://www.bia.gov/service/indigenous-tourism/why-food-sovereignty-matters.

17   Michael Pollan, "How Change is Going to Come in the Food System," *The Nation*, September 11, 2011.

Chapter 7
Who is Hungry?

It would seem that, quite possibly, the ultimate measure of health in any community might well reside in our ability to stand in awe at what folks have to carry rather than in judgment at how they carry it.
– Gregory Boyle

Who is hungry? Let's start with misperceptions, and a personal story that illustrates how something that appears so wrong at first glance can look quite different after closer inspection. During my site visit to one of the food bank's distribution partners, a Cadillac Escalade (yes, the classic story you hear) drove up to receive food. My mind went immediately into stereotype mode. Why is this lady taking food? She does not need it. My food bank spirit sank. A volunteer approached the open window of the Escalade and asked the lady how she could be helped. She replied, "I brought Agnes, a church member, to get some food. I've driven by before and observed what you do. Agnes is elderly, and her husband is disabled; they don't have a car and need food." Lesson learned; let's all take time to see what's happening.

This chapter is the book's heart because it's all about the people—our neighbors facing hunger—and their real-life experiences. They are the people served by the Feeding America network of two hundred food banks and sixty thousand food distribution partners such as pantries, shelters, and after-school programs. Many people harbor stereotypes and misperceptions about our neighbors facing hunger, about who is hungry, and about the different types of poverty tied to hunger.

In the United States, thirty-four million people are food insecure, which means they lack access, at times, to enough food for an active, healthy life for all household members and experience limited or uncertain availability of nutritionally adequate foods.[1] Nine million of that number are children. People of color are disproportionately impacted, older adults, mentally and physically challenged neighbors and too many working households. Lack of adequate income is the primary cause of food insecurity. Hunger and poverty rates tend to be similar. For example, the poverty rate in 2001 was 11.7 percent.[2] The food insecurity rate was 10.7 percent.[3] In 2019, both rates matched at 10.5 percent.[4] Over the past twenty years, each trend has been relatively flat; however, population growth translates into more people facing hunger. Two significant exceptions to the flat trend line were the four to five years after the Great Recession in 2007, when it took some people in poverty time to right their lives again. The second was a massive spike for several months when COVID-19 hit. A survey by the Census Bureau in 2020 revealed that 23 percent of households were food insecure.[5] In 2023, charitable food distribution in many areas remained above pre-pandemic levels due to high inflation, unaffordable rentals and homes, high food costs, and low wages. As Michael Pollan stated about the pandemic's effects on hunger and poverty, "For our society, the COVID-19 pandemic represents an ebb tide

1   "Hunger Facts," Feeding America, 2023,, https://www.feedingamerica.org/hunger-in-america.

2   Bernadette D. Proctor and Joseph Dalaker, "Poverty in the United States: 2001," United States Census Bureau, September 2002, https://www.census.gov/library/publications/2002/demo/p60-219.html.

3   Mark Nord, Margaret Andrews, and Steven Carlson, "ERS Household Food Security in the United States, 2001," USDA, October 2002, https://www.ers.usda.gov/publications/pub-details/?pubid=46684.

4   "A Profile of the Working Poor, 2019," report 1093, Bureau of Labor Statistics, May 2021, https://www.bls.gov/opub/reports/working-poor/2019/home.htm.

5   Rachel Zimmer et al., "Exploring the Perceptions of a Fresh Food Prescription Program during COVID-19," *International Journal of Environmental Research and Public Health* 19, no. 17 (September 2022).

of historic proportions, one that is laying bare vulnerabilities and inequities that in normal times have gone undiscovered."[6]

Before I go further, here's how poverty is defined economically. The federal poverty level for a family of four is $30,000 annually.[7] It varies state by state to reflect those areas' higher or lower living costs. Many folks at 150 percent of the poverty level struggle to make ends meet. Many parents work multiple jobs. While writing this book, I hired a young man who was starting his landscaping business to cut my lawn. He is working three jobs to get his feet under him and feed his family. The impoverished struggle to think of the future because they focus on surviving for the next few days or weeks. In this mindset, neither adults nor children consider college, careers, or higher achievements. Even if they do, they often feel that these dreams are unattainable and that their lot in life is just to try to survive.

So, let's look closer: who's hungry? Those guys at the end of the exit ramp or at the traffic light holding the cardboard sign that says, "Need help, will work for food"? Based on my experience over the years, although the homeless are the most visible, they represent the minority of food relief recipients. The remaining majority include children, seniors, and the disabled. They are mostly the working poor or are under employed, and many are single mothers. People of color are disproportionately affected by both poverty and hunger. Most of these people did not make bad decisions; life happened to them. Many hungry seniors are veterans who served the United States, with many still suffering from wartime injuries. In 2014, the Second Harvest Food Bank of Central Florida and the other two hundred members of Feeding America conducted a nationwide hunger study. At that time, 57 percent of the households served by food bank pantries had to choose between paying for food and the rent or mortgage at least once in the past year; 70 percent had to choose between

---

6   Michael Pollan, "The Sickness in Our Food Supply," *New York Review of Books,* May 12, 2020.

7   "Poverty Guidelines," 2023 Poverty Guidelines: 48 Contiguous States, ASPE, Office of the Assistant Secretary for Planning and Evaluation, U.S. Department of Health and Human Services, https://aspe.hhs.gov/topics/poverty-economic-mobility/poverty-guidelines.

paying for utilities or buying food; and 66 percent were forced to choose between food and medical care.[8]

A sizable portion of the hungry in the United States are, ironically, the very people who harvest our crops and help put food on our tables. Sr. Ann Kendrick and Felipe Sousa-Lazaballet of the Community Hope Center in Apopka, Florida, state that "the foreign-born population pays $3.5 billion in taxes. We enjoy the economic opportunity they have made possible. They are not part of the problem; on the contrary, they are part of the solution."[9] Gary Tester, CEO of Catholic Charities in Central Florida, explained the challenges facing immigrants that often push them into poverty and hunger. These people are leaving their home countries, often so their families can escape economic collapse, food shortages, dictatorial leaders, and spreading violence. Thus, they arrive with few resources. They want to work and are lining up for work authorizations that previously took three to four months, but now are taking twelve to eighteen months. The difficulty of finding work—especially well-paid work—combined with the high price of housing, propels many immigrants into poverty and hunger.

Generally, poverty can be divided into two types: situational and generational. Situational examples include poverty generated by events such as the Great Recession, temporary job loss, long-term illness, divorce, or death. Such disasters can affect us all regardless of our income or lack thereof. The following letter, received at the food bank from a household in need, is an example of situational poverty:

> I am handicapped, and my husband is the only one working for a family of four. He makes a meager $12 a month over the limit to receive food stamps, and I can

8   Nancy Weinfield et al., "Hunger in America 2014: A National Report," Feeding America, August 2014, https://www.feedingamerica.org/sites/default/files/2020-02/hunger-in-america-2014-full-report.pdf.

9   "From One Immigrant to Another: Words of Encouragement," Hope Community Center, July 10, 2023, https://www.hcc-offm.org/from-one-immigrant-to-another-words-of-encouragement-from-hopes-executive-director/.

no longer work. I await an answer from disability, due in the next four months. I was recently injured due to a drunk driver while five and a half months pregnant and had my four-year-old in the back car seat. She now has Tourette's and takes medication. I have had five surgeries and suffered permanent nerve damage and brain injury. I have titanium rods in my neck and back. I suffer from muscular-skeletal diseases. My husband pays for our insurance and drives one and a half hours to work daily. Not to cry on your shoulder, but it is tough because we have depleted our savings accounts and live paycheck to paycheck. I waited five years for a settlement from the accident, and the attorney got more than I did. I am happy to be alive for my two little girls. Medications, car, house, and utilities have burdened my husband while we await an answer. Tell me what you must do to get food in our home.

Generational poverty is when a family lives in poverty for at least two generations. Financial issues cause such poverty, but then it is amplified by feelings of hopelessness and continual stress. Urban Ventures, a poverty-fighting organization focusing on education, says, "Hopelessness is the key factor in creating the cycle—one generation to the next. Without hope and the belief that life can be better, the motivation and energy needed to break the cycle are very low."[10]

The story that follows exemplifies generational poverty and the challenge and grit required to rise above. I interviewed Barbie Izquierdo, a single mom with two children who grew up in poverty in North Philadelphia and was a victim of abuse. She courageously told me about part of her journey through life. She said it wasn't until she became a teen mom that she realized all these systemic barriers in place that seem to keep people like her who have so much potential away from opportunities, away from resources, and just stuck in this

---

10   "Lack of Resources Cause Underdeveloped Learning Generationally," Nasdaq, November 1, 2022, https://www.nasdaq.com/articles/lack-of-resources-cause-underdeveloped-learning-generationally.

cycle of poverty and shame. An example was when she got a slight raise of less than a dollar an hour. She discovered it would put her income slightly over the requirement to qualify for the food stamps she received. So, if she accepted the raise, she would lose so much more income—thousands of dollars. One step forward, two steps back. She fell off what's called the "benefits cliff." She told me, "As a single mother, I'm the sole provider for the kids and supposed to keep them safe and provide the necessities and, at times, could not. Tremendous guilt and shame as a result." Barbie played by all the rules, yet the challenges she faced were great.

Eventually, Barbie had the good fortune to be asked to tell her story for the documentary *A Place at the Table*, which gave insight into the hunger problem in America. She became involved with Witnesses to Hunger, a group of women who share their stories with various groups and decision-makers and advocate for women in poverty across the nation. Through these experiences, she discovered new talents, such as public speaking, data management, engaging with the media, building relationships, writing, blogging, working as an organizer and advocate, and fighting for the benefit of other people. She is now an advocacy director with Feeding America. She is leading an effort to engage people with lived experience to speak for themselves, educate others on what living in poverty is like, and create better solutions and policies. Barbie says, "No matter how misunderstood, undervalued, or unappreciated you feel, the shame starts to fade when you accept your pain as your strength. When you understand that the moments during which you felt like everything around you was trying to break you happened so that you could be a source of strength—for yourself, those around you, or for a stranger. I am still searching for my tribe of *she*roes to build this sisterhood and help support and empower each other through this work that is so vital right now."[11] Barbie is one example of thousands of people across America with passion, skills, and talent who just need that one small break to show what they're capable of and how they are such an asset to society.

---

11   Barbie Izquierdo, "Barbie Izquierdo: I See You, Mujer Ponderosa," Feeding America, March 24, 2022, https://feedingamericaaction. org/blog/barbie-izquierdo/.

Part of the difficulty of escaping poverty is that it is expensive to be poor in the United States. For example, Shelia, a single mother, didn't think her electricity would shut off because she was only $80 behind. She was planning to pay it off in a future paycheck. When it was cut off, she called the utility company and paid off the bill with what was supposed to be rent money. Then she was asked for a "new customer" deposit of over $200, which she did not have. Thus, the lack of $80 ended up leaving her without electricity and needing to pay more than double what she had originally owed.

Basic necessities take a greater percentage of poor people's income. United Way conducted a study of working households above the poverty level in a variety of states. Results showed that 42 percent of working households struggled to afford rent, buy food, or pay for utilities. They could not afford the basics. Most families did not have $400 in a checking account for emergencies. While these households are not earning enough to afford essentials, their earnings are often too high to qualify for assistance.[12] Most poor people rent their homes because they cannot afford the down payment necessary for a mortgage. Unfortunately, rent is usually much higher than mortgage payments, so one in four renters spend more than half of their income on housing.[13] In most markets, a minimum-wage worker must work more than one hundred hours weekly to pay the rent.[14] Another barrier for the poor is the expense of childcare. Childcare costs in the United States range from about $5,500 in Mississippi to around $24,200 in Washington, DC, annually. A mother earning minimum wage in the capital would have to work for forty hours a week for forty weeks to be able to afford day care for one infant.[15]

---

12   Unitedforalice.org. 2023, National ALICE Research Advisory Committee

13   Cbap.org/research/housing/renters. March 9, 2017

14   Adam Barnes, "You have to work more than 100 hours a week to afford a two-bedroom rental on minimum wage: Report," *The Hill*, June 15, 2023, https://thehill.com/business/4052150-you-have-to-work-over-100-hours-a-week-to-afford-a-one-bedroom-rental-on-minimum-wage/.

15   "Child Care Costs in the United States," Economic Policy Institute, updated October 2020, https://www.epi.org/child-care-costs-in-the-united-states/.

While basic living costs take more of their paycheck, the poor also spend more money than average Americans in absolute terms. Many people without sufficient cash flow rent to own for significant expenses, such as appliances. Over time, what they pay in monthly fees amounts to an exorbitant final cost, much more than if they had the cash to buy the item up front. It's financial quicksand. Similarly, the poor are often excluded from the savings of buying in bulk. Bulk purchases are impossible when you don't have transportation or an apartment with adequate storage space. Similarly, buying quality over quantity is not an option for a low-income household. The rich can afford to purchase well-made boots that last for years. People experiencing poverty make do with the poorly made boots they can afford, which have to be replaced often. The poor buy cheap used cars that must be fixed regularly, often paying high interest because they do not have the entire cost to hand. Calculated out, the rich often spend less on consumer goods because they can afford quality, which lasts and doesn't need expensive maintenance.[16]

Poor people also pay more in taxes and financial fees than other Americans. Regressive taxes in which everyone is charged the same amount, such as sales taxes and user fees, cost low-income people a larger share of their income than their middle- and high-income peers. "According to the study, in 2015, the poorest fifth of Americans will pay on average 10.9 percent of their income in state and local taxes, the middle fifth will pay 9.4 percent, and the top 1 percent will average 5.4 percent."[17] Also, many low-income people are unbanked (not served by a financial institution) and thus are nearly eaten alive by exorbitant fees. Unbanked customers spend approximately 4 to 5 percent of a payroll check to cash it. A household with a net income of $20,000 may pay as much as $1,200 annually for alternative service fees.

---

16   Candice Elliott, "The High Cost of Being Poor," Listen Money Matters, 2023, https://www.listenmoneymatters.com/the-high-cost-of-being-poor/.

17   Patricia Cohen, "Study Finds Local Taxes Hit Lower Wage Earners Harder," *New York Times*, January 13, 2015.

Time also handicaps the poor and worsens their situation. They often cannot afford to live close to their workplace. Indeed, they are frequently faced with a long commute that prevents them from doing many other things, such as working a second job, helping kids with homework, and preparing dinner. Inconsistent and often inflexible work schedules are also a burden. Medical appointments, teacher meetings, and other essential trips can be impossible in some work situations. A consistent schedule makes all those things more accessible.

So, let's do the math—there's too much month at the end of the money. With half of one's income going to housing and a quarter going to food, it doesn't take much to realize how expensive it is to be poor in this country and how hard it is to break out of the cycle of poverty. And this doesn't include childcare. If someone is working for the federal minimum wage of $7.25 per hour, the odds are they won't be able to afford a car or public transportation.[18] Here's an example based on a family of four with two adults working at $11.00 an hour. This simple illustration does not include daycare costs, which can be exorbitant, a clothing budget, entertainment, or any extras, and assumes the beat-up car is paid off. The household's monthly income (based on a four-week month) is $3,520. Their income is taxed, and their net monthly pay is $2,155.[19] Their expenses include rent at $1,700, car repairs and gas at $280, phone at $140, utilities at $200, food at $750, and healthcare costs at $550—those total $3,650. That yields a monthly deficit of $1,495! Is it any wonder people are committing crimes, selling drugs as a side hustle, and rioting in the streets? I'm not supporting those activities, but understanding poverty can explain some of the street's frustration.

My wife and I did the SNAP Challenge for one week to better understand what a shortage of money could feel like. This was far from enduring the reality. It provided us a brief snapshot, not the experience of a chronic situation. SNAP,

---

18   "Minimum Wage," Department of Labor, USA.gov, updated
        December 6, 2023, https://www.usa.gov/minimum-wage.
19   If you make $30,000 a year living in the region of Florida, you will
        be taxed $4,136. That means your net pay is $25,865 per year or
        $2,155 per month. Your average tax rate is 13.8 percent, and your
        marginal tax rate is 19.7 percent.

formerly called food stamps, is a government nutrition program that provides a small amount of financial assistance to low-income people. So, my wife and I lived for a week on what a couple gets from SNAP—$86! This worked out to about $2 per meal. I gained a better appreciation for what it's like to go to the grocery store while poor. I had to watch every cent, worry about getting to the checkout without enough money, struggle to eat healthy when the cheap choices were packed with salt and carbohydrates, and skip the allergy medicine that would have eaten up a portion of our budget. To get through the week, we had to put less ground beef in the chili and buy fewer tomatoes and slice them thinner. And then there was the stigma of being poor, if only for a week. As my wife and I discussed our choices in the grocery aisles, I realized people might overhear and think we needed more. I lowered my voice and had to control myself when I saw one of our much-needed staples at half price.

Have friends over for dinner? Enjoy the occasional drink? Eat a variety of foods? Forget it. I cannot fathom living like this for an extended time. Our experience was brief, and we knew we would soon be back to our everyday lifestyle. Also, we started the experience with food already in our pantry. Fortunately, we had a car to get to the grocery store—40 percent of SNAP households do not have access to reliable transpiration. And we were not dealing with childcare expenses, debt, back rent, domestic issues, or other adverse life events. Nevertheless, the experience opened our eyes, and our empathy grew. Being comfortable can cause indifference. Being secure can make us unresponsive.

Here's a broader example that provides another lens to better understand another's walk in life. While hosting a group of forty professionals from across the country a few years ago, I had them participate in the Cost of Poverty Experience (COPE).[20] It's an immersive two-and-a-half-hour exercise where participants are assigned roles such as a low-income single mom with three kids, a senior citizen

20   "COPE: A Think Tank Experience," Think Tank: Rethinking Poverty," 2023,, https://thinktank-inc.org/cope.

living off social security, a pawn shop operator, a jailer, a health clinic worker, a day care operator, and many others. The aim is to make ends meet and provide for yourself and your family amid many challenges in a chaotic environment. I failed to provide for my family during this experience while trying to abide by the rules. There were so many obstacles, and I needed more time and resources. The opportunity to put yourself in someone else's shoes, even if only for an exercise, is eye opening and can be a life-changing experience. Some participants took this experience back to their communities and had groups there participate in COPE because it had been such a powerful wake-up call for them.

Over the years, I have heard or witnessed many negative perceptions of low-income people. These attitudes are destructive, alienating, and ill-informed. An article in the *Philadelphia Inquirer* reported, "Many Americans disdain the poor—and science proves it. When people were placed in neuroimaging machines and shown photos of the poor and homeless, their brains responded as though they depicted things not human—a sign of revulsion."[21] Once you've dehumanized a person, it's easier to neglect them. Some negative perceptions are based on the idea that the poor commit fraud to obtain access to government services, such as food stamps. Is there fraud? Of course—there will be scammers and cheaters with any program of that scale. The media sometimes loves to cherry-pick and sensationalize a food stamp story. More than 50 percent of Americans believe that most users abuse SNAP, lying about their situation to gain eligibility. However, the latest statistics indicate that SNAP payments are accurate over 92 percent of the time, while less than 2 percent of food stamps are sold illegally.[22]

Another misconception many people have is that their accomplishments and success are solely due to their effort. This can translate into a belief that they are more deserving than others. They often ignore that they received help over

---

21   Alfred Lubrano, "Reacting to the Poor—Negatively," *Philadelphia Inquirer,* August 5, 2013.

22   "Welfare Fraud Statistics," Balancing Everything, May 20, 2023, https://balancingeverything.com/welfare-fraud-statistics/.

the years from others, such as family, friends, and colleagues. Can they—can any of us—say we did it all on our own?

Indeed, I have found that many low-income people have a strength that many well-off people lack. I am constantly amazed and humbled by their stories of struggle and survival. Here's the story of Bonita, a young lady who grew up in poverty; she didn't have a choice of what zip code she was born into or her family environment. She isn't a moral failure; just in circumstances beyond her control, and she overcame unbelievable odds against her. Her story is one of millions. It's the real world of our low-income neighbors.

Bonita grew up in the rural South, surrounded by generational poverty; her mom was a victim of domestic abuse and worked two jobs. Her dad was constantly in and out of jail, a failure as a father. As a young child, she liked to hang out at her grandfather's pool hall; he was her only sense of a father. It was a safe place because the drug dealers looked after the kids. This was "normal" behavior in that small town.

As high school graduation neared, she worked at McDonald's paying for her senior portraits. After graduation, she wondered: what now? She had no idea what to do, and no source of guidance. As she tried to figure out her next steps, she worked two jobs—one at 7Eleven and the other at Sears. Her job hours were 10:00 p.m. to 6:00 a.m. at the first job, then 9:00 a.m. to 3:00 p.m. at the second. She wanted to go back to school, but as a low-income single mom, this was not possible. Her youngest child was born with Down Syndrome and required heart surgery at three months old, which imposed an incredible financial hardship. She remembers using a camping light to read at night because she couldn't afford the utilities. Syrup sandwiches were not uncommon for dinner. She bought three of the twelve-packs of ramen noodles instead of lettuce, tomato, broccoli, and bananas. She told her two boys not to sign up for school sports until she could re-qualify for SNAP. They needed food to compete. Bonita can remember many times when she didn't eat but ensured her kids did. Grandma came in to watch the kids; childcare was unaffordable. Her employer wanted to give her a twenty-five-cent raise, but she politely declined because it would put her income a few dollars over the SNAP qualification. If she accepted the raise, she would lose $200–$300 monthly in SNAP, falling off a financial cliff despite a meager raise. This situation forces people who want to better themselves into limbo. The law of unintended consequences, a disincentive to work!

She eventually found a church home that focused on helping neighbors in need and joined their Circles program. Circles USA is an organization that "gathers middle-income and high-income volunteers to support families in poverty. Surrounded by people who have landed jobs, negotiated a lease, or managed credit card debt, for example, people experiencing poverty are more equipped to achieve long-term financial stability."[23] This was a breakthrough for Bonita. She had never realized she possessed many skills; just didn't know the names of those skills or the right way to use them. She now facilitates Circles at her church home. Bonita has a decent job that she loves and is in the process of buying her own home.

Today, food banks are placing more emphasis on understanding poverty and the poor. Claire Babineaux-Fontenot, CEO of Feeding America, is leading a significant initiative to center the people that member food banks serve. In 2022, she asked Barbie Izquierdo to plan a national conference titled "Elevating Voices." This conference was designed by people with lived experience of poverty and hunger. It incorporated their perspectives, providing increased dignity and efficacy to programs and services for people facing hunger. I regard this inclusion of neighbors' voices as another significant milestone in Feeding America's history. The historical approach has involved the food distribution partners, such as pantries and food banks, whereas the pantries work directly with the neighbors. This new movement will result in ever-deeper levels of partnership within and across the charitable food system, all with the benefit of better service to neighbors. A higher level of inclusion, equality, and dignity is the goal.

I have deep respect for people like Barbie, Bonita, and thousands of others who find the strength and fortitude to

---

23   "Building Community to End Poverty," Circles USA, 2023
       https://www.circlesusa.org/#:~:text=WE%20REDUCE%20
       POVERTY,achieve%20long%2Dterm%20financial%20stability.

improve their circumstances against incredible odds. I wonder how those of us who have not had those challenges would fare. There is an exhibit in the back of this book titled "Could you survive in poverty?"[24] I encourage you to take the test. The results can be enlightening.

This chapter closes with a story from a neighbor. So many people are in situations that life dealt them; they aren't in that position due to bad decisions or lack of a work ethic:

> Hi, my name is Tom Smith; I am married and have two teenage girls who are always hungry.[25] We have tried to get food stamps but have been denied. I make a modest amount of money working for (the local government) but after rent, medical bills, utilities, gas, phone, etc., I barely have anything left each month for food for my family, and my girls always complain about being hungry. We are new to this area after losing our home to foreclosure, and my wife can't drive or work due to a disability, which has also been on appeal for two years. I wonder if I can get help for my family nearby to assist with a decent Thanksgiving meal. At this point, I am looking at turkey sandwiches . . . which is tough to explain to my pre-teens. We could use help with healthcare. We pay out of pocket, and my wife has seven medications and life-threatening seizures if she doesn't take them. We just had an emergency visit to the hospital last week, and I have no idea what the bill will cost us. We are buried, and I don't know if we can make it. Any help you could offer or point me in the right direction would be greatly appreciated. Thank you, and God Bless.

---

24   From Ruby K. Payne, *A Framework for Understanding Poverty*, 4th rev. ed. (aha Process, Inc.: 2005). Reprinted with permission.

25   Name changed for anonymity.

Chapter 8
Friends of the Food Bank

*It is in your hands to create a better world
for all who live in it.*
Nelson Mandela

In this chapter, I want to identify the generous people who make our nonprofit missions possible and their reasons for doing so. I will include various categories of "friends of the food bank," such as donors of financial resources, food, and time. Two other important categories are the media and the multiple celebrities who add their star power to the cause. Donors vary dramatically in why they give and how they give. No matter the donation size, they all add up to make a collective impact on our world.

Financial Donations
A wide variety of entities donate to food banks, including individuals, businesses of all sizes, faith-based groups, philanthropists, private foundations, educational institutions, and the government. They range from elementary school kids donating pennies and nickels to billionaires sharing their riches. Giving USA reported that Americans provided $484.5 billion in 2021—67 percent from individuals, 19 percent from foundations, 9 percent in bequests, and 4 percent from

corporations.[1] This generosity reaches many recipients, such as faith-based organizations, education, human services, private foundations, the arts, and environmental causes.

I'll start with a story that I'll never forget. One afternoon at the food bank, I was opening the mail when I came across a tiny pale blue envelope with my name handwritten on it in shaky cursive. I opened it, and inside was a matching robin's-egg-blue piece of stationery that smelled like lilac perfume. Also enclosed was a one-dollar bill and a note from a senior citizen. She explained she was a widow, and her wish, with her meager resources, was to help feed kids. The magnitude of that donation hit me hard. Great attention is typically paid to the most significant contributions. This one in the pale blue envelope had as much or more impact than some of the most significant donations we had received because of its incredibly personal, heartfelt nature. I felt that a one-dollar gift had a million dollars of responsibility. That donation was a powerful reminder of people's trust in an organization to do the right thing. Every dollar counts, and everyone makes a difference. Recognizing the impact of every dollar is tied directly to one of the values of the Second Harvest Food Bank of Central Florida: stewardship of resources. Occasionally, I reminded staff, the board, and myself that these donations are not ours; they belong to the community. Every gift makes a ripple like a pebble thrown into a pond. Throw enough rocks in, and lifesaving waves reach the shoreline.

One of the most important jobs in nonprofit organizations is raising funds. My initial background in the advertising business prepared me well because knowing how to communicate effectively is vital to raising resources and revenue. What I learned somewhere along the line was, "It's not about you." That is counterintuitive at first glance. Why wouldn't a nonprofit leader tout the many fine qualities of their organization? I'm not saying don't do that, but one's primary attention must include the donor's desires and goals. Joyaux Associates, a consultant organization for nonprofits, talks about donor-centrism, "which

---

1   "$484.85 Billion," Giving USA, https://givingusa.org/wp-content/uploads/2022/06/GivingUSA2022_Infographic.pdf.

is another way of saying building trust. . . . Trust that donors play an essential, vital, central role in your mission's success. Trust that your organization does worthwhile things with donor gifts. Trust that your organization conducts its operations efficiently. Sadly, most organizations focus on their needs and why their good work requires donations."[2] It's critical that you have a donor perspective.

I learned a lot about what donors are looking for while serving for several years on a panel that chose winners for Bank of America's Central Florida "Neighborhood Grants" initiative. We reviewed several finalists' applications, scored them, and agreed on the final winners. I was fortunate to be the only nonprofit organization in the group, listening to deliberations among major corporate donors and foundation executives. I gained such insight from hearing what they deemed necessary for a successful application! I would always return to my office with the key points and share them with the team. The panel members were looking for organizations with strong leadership and a solid track record. They wanted to see the impact of their investment and, ideally, systemic solutions being put forward in response to problems. They were interested in innovative and effective solutions.

Donors also want to see organizations working together, combining forces, and performing strategically. Many times, if nonprofits have not demonstrated that they are working collaboratively, they miss out on funding. Patty Maddox, former president, and CEO of the Winter Park Health Foundation in Florida, points out that, oftentimes, "Nonprofits are literally within a half mile of each other and don't know the other even exists." At times, the foundation saw nonprofits fail to remove their barriers to working together. As a result, the foundation would walk away. An African proverb hits home regarding collaboration: "If you want to go fast, go alone. If you want to go far, go together." One of the many strengths of the Feeding America national network is that its members are collaborators on a massive

2   Simone P. Joyaux, "A Donor-Centered Organization, Your Donors, & Relationship Building," Joyaux Associates, 2023,, https://www. simonejoyaux.com/downloads/ArticleDonorsDonorCentrism.pdf.

scale: two hundred food banks partner with sixty thousand charitable food distribution sites. In addition, countless local, regional, and national partnerships exist. And those collaborations continue to multiply daily.

Individuals give for a variety of reasons. A 1994 study analyzed the motivations of individuals relative to their interest in and support of nonprofit organizations. It identified seven different types of individual donors: "the devout, investors, socialites, altruists, re-payers, dynasts, and communitarians, folks who want to make their communities a better place to live and work."[3] Another way to think about how people donate is whether it's through their hearts or their heads. Some folks rely heavily on the emotional side and are empathetic. Others want numbers, stats, and facts. A nonprofit leader must speak to a specific donor's mindset and where they are coming from. And, yes, there are many times somebody is using both heart and head. The critical point is to get to know the person and what motivates them. Usually, the more significant the amount of money involved, the more time it will take. Build the relationship; build the trust first.

I recall connecting with a high-profile philanthropist and being tempted to ask for money immediately. Taking a pause with a longer-term approach to building the relationship and understanding what was important to him. I gradually engaged him in our mission by tapping into his love of cooking. Six to eight months later, I met with him on our capital campaign and asked if he would consider becoming involved in reaching the goal. I paused, waiting for his response; ten seconds sometimes seemed like an eternity. He said, "Sure, I'll commit to *XYZ* dollars." It was twice what I was going to ask for. The point of sharing this has to do with what he said after making his commitment. He said, "Dave, thank you for taking the time to get to know me before asking for anything. So many groups come to me and ask for money and know nothing about me."

---

3   Russ Alan Prince and Karen Maru File, *The Seven Faces of Philanthropy: A New Approach to Cultivating Major Donors* (Jossey-Bass, 1994); Russ Alan Prince, "The Seven Faces of Philanthropy (Why the Wealthy Give to Charity), *Forbes*, March 18, 2016.

Similarly, shortly after COVID-19 hit, I received a call from a colleague in another state. He told me he had recommended our organization to a major company looking for the go-to place to invest in food relief in Central Florida. I then received a call from the donor, Kroger. The company's representatives told me they would donate nearly $1 million to the food bank for COVID-19 relief efforts. I asked the Kroger executive about their specific requirements for the gift, and their response was, "You know best how to spend this money; you have total freedom in what you wish to do." Such relationships can be golden because the likelihood of receiving additional funds is enhanced once you have demonstrated the meaningful impact of the initial donation. Thus, rather than chasing every dollar, the money will start to follow you.

It is also important to know your audience and what motivates them. Once, I was preparing to visit our state capital to receive funding for one of our programs. While doing my prep work for the visit, I spoke to a lobbyist and asked what the giving climate was like at that time. His response was, "Empathy is dead." Consequently, I changed my initial narrative. I made a presentation regarding a return on investment for government funds—an investment in a trained workforce that would become taxpaying citizens. I showed how state financial investment would generate several jobs with decent wages and yield a successful graduation and job placement rate. By effectively addressing my audience's concerns, I received double the amount I requested.

Donor relations don't end after receiving the gift. It's important to thank donors and keep them informed. Unfortunately, over the years, I have heard too many times from donors that organizations fail to say, "thank you." I recommend setting up a system to manage how donors will be thanked. In addition to the leader of the nonprofit, other appropriate staff or the board should be involved. Ideally, have someone who benefits from the organization's mission say thank you when possible. Use acknowledgment letters, emails, and personal meetings, depending on the situation. Donors at the Second Harvest Food Bank receive an acknowledgment letter no matter the donation amount. I would sign all my "thank you" letters by hand;

there are some things you don't automate. It was a meditative, prayerful exercise that kept me humble and grateful for the donors' trust in the organization. It was a great reminder that it's not our money, it's the community's, and they trust and believe in us to do the right thing.

Equally, it is important to keep the donor informed. Start an ongoing relationship. I made it a practice to send major donors a short personal email from time to time about what was happening with their goodwill, often as a story "from the road." I called them "Sixty Second Snippets." Carrie Morgridge from the Morgridge Family Foundation appreciated these. She said, "Those snippets made us feel in the moment, made the members of the foundation feel like they were part of the journey and part of the solution."

Food Donations

Along with financial donors, another vital source of support for food banks is food donors. These come in many forms and sizes, ranging from farms to convenience stores and all sorts in between. In the 1980s, manufacturers such as Pillsbury, Kraft, General Mills, and others were the primary sources of donations. Today this has shifted to grocery stores, farms, and the USDA.

A lot of excess food exists that could be made available for donation. A 2012 Natural Resources Defense Council issue paper stated that 40 percent of food grown in the United States goes uneaten.[4] Losses in our food system occurs throughout the supply chain. Food is lost on farms; during processing, distribution, and storage; in retail stores and food service operations; and in households, with the reasons varying at each stage. Food manufacturers may make labeling mistakes, and as a result, the product cannot be sold at retail. Defective products and packaging result in waste. On average, fifteen thousand new food products are launched yearly, and only about 2 percent are successful. At the retail stage, there are

---

4   Dana Gunders, "Wasted: How America Is Losing up to 40 Percent of Its Food from Farm to Fork to Landfill," NDRC issue paper, August 2012, https://www.nrdc.org/sites/default/files/wasted-food-IP.pdf.

various reasons for waste, including the expectation of cosmetic perfection, pack sizes that are too large, expired sell-by dates, damaged goods, outdated promotional products, and unwanted items. Foodservice (restaurant) surplus occurs due to unfinished meals, large portion sizes, kitchen loss, and menu changes. For consumers, confusion over label dates, spoilage, bulk purchases, poor planning, and over-preparation lead to food not being consumed.

One way the food industry reduces waste is through donations. In 2022, these donations were enough to provide 5.2 billion meals to Feeding America and its network of two hundred food banks.[5] This includes food from specific USDA programs such as TEFAP. Food banks receive products from the entire food supply industry. These contributors include farmers, processors, manufacturers, wholesalers, retailers, and food service establishments. These food donors donate for a variety of reasons and derive various benefits. Most companies receive a tax deduction based on the value of the food they donate. At the same time, they save money by reducing their dumping costs. Also, they are protected from legal liability under the Good Samaritan Act, and food banks pick up the food, eliminating potential delivery costs. Feeding America members uphold food safety standards, and the companies get to participate in solving hunger in our country.

One of the most significant providers of food to various aid programs is the USDA. To ensure a minimum income to farmers, the USDA regularly purchases surplus food, which it then channels for our neighbors facing hunger through various federally funded programs, including TEFAP; School Lunch and Breakfast programs; the Women, Infants, and Children (WIC) program; SNAP; the Child and Adult Care Food Program; the Commodity Supplemental Food Program; the Farmers Market Nutrition Program; the Farm to School program; the Food and Nutrition Service's disaster assistance; the Food Distribution Program on Indian Reservations; and

---

5   "About Feeding America," Feeding America, 2023, https://www. feedingamerica.org/about-us/press-room/mealconnect-5-bil-lion-pounds-rescued#:~:text=Feeding%20America%C2%AE%20 is%20the,people%20in%20need%20last%20year.

the Fresh Fruit and Vegetable Program. In addition, food banks have been the recipients of surplus food left over from wars. During the Gulf War (1990–1991), $300 million worth of rations were distributed across the country. The initiative was called "Operation Desert Share."[6]

Along with the USDA, food manufacturers were among the earliest contributors to food banks. Some, such as Kraft, Nabisco, General Mills, and Beatrice Foods, were on board in the 1980s during the development of Feeding America. Today, if you look at Feeding America's website, you'll see a "who's who" of names in the industry. Just about every major company is involved to some degree. Many are providing not only food but also financial support and volunteers.

The US food manufacturing industry is a wonder. For a few years, when I was the vice president of business development at Feeding America, I led a team that worked directly with the industry on donations. We had a chance to see this multibillion-dollar industry with thousands of facilities up close. It's almost miraculous to think about the ability to feed a population of 332 million every day, not counting anything that's exported. One sight I'll never forget was during a visit to a salmon cannery in Ketchikan, Alaska, that appeared to be right out of the 1920s. Boats docked right next to the factory, and huge, flexible pipes, three feet in diameter, were connected to the hold of the ship. The other end opened to a fast-moving conveyor belt that carried the catch to the processing tables. It looked like an endless sparkling silver streak. The machinery used to process the fish looked like something out of the steampunk era, incredibly low tech but effective. The cannery was canning salmon and freezing steaks for the food banks from bycatch. These are undersized fish caught accidentally and not allowed for sale. However, they can be donated. This was made possible through Tuck Donnelly, a champion and founder of Northwest Food Strategies. Many other food processing facilities donate to food banks, from one-million-square-foot distribution centers within which a small city could be

---

6   Christine Kotbra, "Food Banks and Jails Share War Surplus," *New York Times*, November 10, 1991.

contained, to a robotic grocery delivery facility that looks like a scene in a sci-fi movie. All of it is appreciated.

Retail operations also donate considerable amounts of food. The US retail food industry is a multibillion-dollar business with about forty thousand stores.[7] Almost every retail grocery store in the country donates food in their local communities. Champions in the fight against hunger for decades include, but are not limited to, H-E-B, Publix, Safeway, Winn-Dixie, Food Lion, Wegmans, Costco, Sam's Club, Meijer, Walmart, and Target. During the 1980s and 1990s, retailers sent all their unsalable products to large reclamation centers for dumping, resale, or donation. Food banks would have only a few locations to pick up from. A significant change occurred when grocery stores closed their reclamation centers and required food banks to pick up donations at individual store locations. With this shift, perishable products were rescued for the first time through retailers. This had a significant impact on the infrastructure of food banks that had little or no refrigeration and no large fleet of refrigerated trucks to ensure food safety. With the support of monetary contributions from the retail sector and private and corporate donors, the Feeding America network acquired fleets of refrigerated trucks to pick up and distribute fresh foods.

Kroger, one of the largest grocery chains in the nation, was the first major company to partner with Feeding America in the early 2000s, largely through the efforts of Lynn Marmer, the group vice president for corporate affairs for Kroger. Denise Osterhues, the president of the Kroger Co. Zero Hunger/Zero Waste Foundation and the sustainability and social impact director, spoke to me of their mission to help create communities free of hunger and waste. Kroger wants to transform surplus food into healthy meals, convene thought leaders to advance equitable food access, and offer easy ways for others to join this journey. Throughout our conversation,

---

7   "JICC Calls for Industry Comment on Retirement of Coupon Barcodes," FMI The Food Industry Association, April 29, 2014, https://www.fmi.org/newsroom/news-archive/view/2014/04/29/jicc-calls-for-industry-comment-on-retirement-of-coupon-barcodes.

it was clear that Kroger's commitment was in its corporate DNA. In 2021, Kroger provided donations totaling $181 million to hunger relief, food systems work, and urgent community response.[8] Also, many of its five hundred thousand associates volunteer at food banks and pantries throughout the year. Denise claims that Kroger's impact would not be possible without their partnership with Feeding America because of the nonprofit's efficiency. Kroger works closely with Feeding America members to increase the number of pickups, to make sure the routes to stores are mapped out efficiently, to encourage stores to donate, and to shift the balance to include more fresh produce and other nutritious items. She added, "We will not end hunger until we stop throwing food away—35–40 percent of all food is thrown away!" She also stated, "The Zero Waste/Zero Hunger initiative is built on the absurdity of the food system."

Kroger was an early ally for Feeding America; however, since the year 2000, dozens of other retailers nationwide joined the food security movement, ranging from individual one-store operations to regional and national chains. The world's largest company, Walmart, became engaged and continues to have a significant impact on food security. Their "Fight Hunger. Spark Change." program has helped food banks across the United States. In 2021, Walmart and the Walmart Foundation provided more than $1.5 billion annually in cash and in-kind support for various programs and nearly one thousand hours a day of volunteer help.[9]

More recently, companies other than grocery stores have started donating to food banks. I remember the first time I saw a truckload of unsalable items arriving at our food bank receiving docks labeled "Amazon." Some of the donated goods were nonfood items, such as bicycles, baby strollers, small

---

8   The Kroger Co., *Zero Hunger/Zero Waste Foundation 2022 Report*, 2023, https://www.thekrogerco.com/wp-content/uploads/2022/08/Kroger-Co-Zero-Hunger-Zero-Waste-Foundation-2022-Report.pdf.

9   Walmart, *Environmental, Social and Governance Summary Report FY2022*, 2023, , https://corporate.walmart.com/esgreport/media-library/document/walmart-fy2022-esg-summary/_proxyDocument?id=00000182-21ec-d591-afe2-2bfcb4df0000.

appliances, clothes, school supplies, and health and beauty items. These can be distributed to select charities through the Feeding America network. The people these organizations serve can benefit from receiving this merchandise, which eases their household budgets. In addition to food and home goods, Jeff Bezos donated $100 million to Feeding America members shortly after COVID-19 hit the country. Claire Babineaux-Fontenot, the CEO of Feeding America, stated on their website, "This incredibly generous donation was the largest gift in their history. Countless lives will be changed because of his generosity." Fast-food restaurants, as well as Starbucks and others, represent additional categories.

Another category of retailers that partner with Feeding America includes convenience store and gas station chains, including Wawa. Its unique company culture shows through its service and employees. Howard Stoeckel, vice chair of their board, states that it is a "bottom-up" organization that empowers employees and provides them with store-based-decision-making power. When Wawa opened its first store in Florida in 2012, it invited Second Harvest Food Bank of Central Florida to receive its goodwill. Since then, Wawa has provided nearly twenty million pounds of food for hunger relief nationally.[10] Most recently, in 2023, it donated $1 million to the food bank to expand healthy fresh food markets in middle and high schools in Central Florida.[11] Liz Simeone, the senior manager of the Wawa Foundation and Wawa Community Care, told me that feeding kids is so important because it can change a child's trajectory for life. Wawa provides financial support to various causes, with its employees influencing those decisions. She says it's important to Wawa to become part of the local community and, consequently, an enduring company. Overall, retailers are a vital part of food banks' supply chains. Between 2002 and 2022, they donated approximately twenty-four billion pounds of food—the equivalent of about sixteen billion meals.

10   "Wawa Share Food Donation Program," The Wawa Foundation, 2023, https://www.thewawafoundation.org/our-focus-and-impact/food-donation-connection/.

11   Kevin McIntyre, "Wawa Pledges $1 Million to Florida Food Bank," CStoreDecisions, May 23, 2023, https://cstoredecisions.com/2023/05/23/wawa-pledges-1-million-to-florida-food-bank/.

American farmers are another important source of food for the hungry in the United States. They donate millions of tons of fresh and healthy food annually. Farmers find themselves with surplus crops for various reasons, such as adverse market conditions, off-spec sizes (too big, too small, cosmetic blemishes), cheap imports flooding the country, weather damage, or an overly abundant harvest with no market. The examples of rejected produce range from cucumbers that are off-spec for a pickle company to sweet potatoes the size of footballs. Gleaning also helps farmers reduce food waste. Gleaning involves volunteers rescuing surplus or unwanted produce from the fields for distribution to food banks. There are many gleaning organizations across the country. The largest is the Society of St. Andrew. In the first half of 2023 alone, this organization mobilized over ten thousand volunteers to glean the fields of 571 farmers.[12]

Farmers have been involved with food banks since the early years. In the early 1980s, a Central Florida businessman named Dave Pearlman was flying his small plane over nearby farmland one day when he saw an orange carpet in one of the fields below. As he watched, it looked as though the tractors were covering up the carpet with dirt, little by little. Dave's curiosity got the better of him, so he landed his plane in a grassy area next to the field and asked the farmer what he was doing. The farmer explained that when he harvested this field of carrots, he discovered the vegetables had grown the wrong size and color for sale to the retail produce market. Consumers know what they like to buy in the grocery store, and these wouldn't fit the bill. He explained that this was a disappointment but not uncommon, and he was turning the soil back over so the carrots would rot back into the ground to help fertilize it for future crops. When Dave asked why the carrots couldn't be harvested and given away to people who needed food, the farmer chuckled and replied, "Because I would have to harvest them, and box them up, and then haul them, and I'm not made out of money." Dave's discovery of

12   "Reduce Waste. Feed People," Society of St. Andrew, 2023, https://endhunger.org/.

this surplus food opened the food bank's eyes to the potential of securing much more from farmers. Dave Pearlman went on to be a great advocate and supporter of the food bank in Orlando. He helped establish a foundation that has helped provide food for families facing hunger in Central Florida. The pantry at Jewish Family Services in Orlando bears his name.

Yet another way food banks obtain fresh food is by growing it themselves. Food banks have started farms in some rural areas, utilizing volunteers and staff. Although this food is technically not "donated," it serves as an important source in some areas of the country. In 1997, the Regional Food Bank of Northeastern New York founded a 162-acre farm outside of Albany. This farm is funded by private donors and corporations and staffed by volunteers. It has the dual purpose of providing fresh food to the hungry and educating the broader community about agriculture. More recently, in 2013, the Mid-Ohio Collective, another Feeding America member, established a seven-acre urban "smart farm" in Columbus, Ohio. Here, the organization uses technological innovations to grow more crops on less land.

Traditionally, most food donated to food banks is unprepared. However, a significant amount of prepared food is also wasted and represents an opportunity to provide more food for the hungry. Rescuing prepared and perishable food is a specialized category. These food donations are not leftovers from customers' plates. They come from surplus food due to unwanted, old, or discontinued menu items, over-preparation, and unanticipated inventory fluctuation. Food banks nationwide have refrigerated trucks and staff trained in food safety to pick up prepared foods at various restaurants, hotels, catering events, corporate cafeterias, and NFL, MLB, and NBA games. In the early 1990s, the prepared food programs, many of them already associated with food banks, established FoodChain, a network similar in scope to the Second Harvest network. These programs expanded to numerous cities nationwide with financial assistance, logistics planning and support from the UPS Foundation. From fewer than a dozen prepared food programs in the early 1990s, more than 125 programs were established by 1999. In 2000, after several years of negotiation and national strategic partnerships, and with the urging of the food industry, the boards, and memberships of Second Harvest and FoodChain agreed to merge the two organizations, and together they formed America's Second Harvest.[13]

Many national restaurant chains partner with Feeding America to provide millions of pounds of food annually. Some major contributors include Yum! Brands, which owns KFC, Taco Bell, and Pizza Hut, and Darden, which owns restaurants like Olive Garden, Bahama Breeze, Longhorn Steakhouse, Cheddar's, Eddi V's, and the Capital Grille. Even the bigger corporations help out. When Disney closed its theme parks in Orlando at the start of COVID-19, it was left with food slated to feed millions of guests. The food bank in Orlando received 104 semitruck loads from March 2020 to December 2020! The Orlando Second Harvest Food Bank enjoys a special relationship with Disney and the Disney Harvest program, which donates food from its resorts and hotel properties. Universal Orlando is also a strong supporter of the food bank. Both companies provide hundreds of volunteers and significant financial support. They are incredible corporate citizens. Convention centers are another source of support. Feeding America members located in major metropolitan areas that attract large conventions receive donations after mega-size breakfasts, lunches, dinners, or banquets by conventioneers.

In addition to food donations, food banks must often purchase certain foods to round out their inventory and provide additional quantities. Food banks across the country have local relationships with food wholesalers or, in some cases, manufacturers and can benefit from low prices. These are products that would not qualify as donations. The primary provider to Feeding America members is Project Preserve, part of Second Harvest Food Bank of Middle Tennessee in Nashville, which provides low-cost pantry and meal solutions.

---

13   Doug O'Brien, Erinn Staley, Stephanie Uchima, Eleanor Thompson, and Halley Torres Aldeen, "The Charitable Food Assistance System: The Sector's Role in Ending Hunger in America" (2004 Hunger Forum Discussion paper), The UPS Foundation and the Congressional Hunger Center, https://www.hungercenter.org/wp-content/uploads/2012/10/The-Charitable-Food-Assistance-System-Americas-Second-Harvest.pdf. 2023

The Nashville food bank has developed unique relationships with select food manufacturers and buys in bulk at discounted prices. They then can offer those savings to food banks across the country. They provide several food categories, such as disaster boxes, holiday boxes, staple groceries, diabetic boxes, and more. Many of these products are in demand by pantries but are rarely donated, or else not consistently or in significant quantities. During the COVID-19 relief efforts from 2020 to 2022, Feeding America members had to spend more than $100 million on purchased foods due to the extraordinary demand.

Food donations that involve the general public, such as community food drives by families, schools, scout groups, letter carriers, and companies, are popular. It's a very tangible way of helping. It is small as a category of the total donations food banks receive, but these contributions do round out their offerings. Every can of food or box of cereal does make a difference. It is not a cost-effective way to fight hunger, however. Consider that most food banks can distribute approximately ten meals for a dollar due to the support received from the community. Food banks have the logistics, volunteer help, and scale to achieve this efficiency. The cost of that can of food purchased could have gone much further in its impact. When well-intentioned people buy food for a drive, they pay retail prices. That money could be donated and produce a much more significant impact on food security.

Food banks continue to innovate, and some now offer "virtual food drives." Individuals or groups can visit the food bank's website to provide funds and see what they provide in return for their financial contributions. It simplifies the idea of a drive and provides a better experience for the donor. Groups can register online and have competitions; results can be posted daily on many sites. Ultimately, food banks won't discontinue traditional food drives, as they understand it is a unique "hands-on" experience for some.

A big question for food banks in their search for new food sources is, "Can food keep being produced at current or even higher levels into the future? Are we food secure as a planet?" The answer is a mixed bag, and it's very complicated. On the doom and gloom side, a 2018 *TIME Magazine* article stated, "By

2050, with the global population expected to reach 9.8 billion, our food supplies will be under far greater stress. Demand will be 60% higher than it is today, but climate change, urbanization, and soil degradation will have shrunk the availability of arable land, according to the World Economic Forum. Add water shortages, pollution, and worsening inequality into the mix, and the implications are stark."[14] According to the USDA, the number of farms in the United States peaked at 6.8 million in 1935. In a recent survey, there were 2 million farms in 2022. Similarly, the acreage of farms continues its downward trend. Fortunately, some of that downsizing has resulted from technology and scientific advancements that produce much higher yields in smaller areas.[15] On the bright side, "since 1960, the world's population has more than doubled—yet global food production has tripled while using only 15% more land. . . . 'Farmers are wildly more sustainable now than in 1980,' says Jack Bobo, JD, director of global food and water policy at the Nature Conservancy. 'It's not as if things are bad and getting worse—they're good and getting better, just not fast enough. Historical trends won't get where we need to go by 2050; We need to do things differently.'"[16] For this book, I interviewed Eric Bost, who previously served as the undersecretary for Food, Nutrition, and Consumer Services at the USDA from 2001–2006 under President George W. Bush. He is now the deputy director at Borlaug Institute for International Agriculture and Development in the Texas A&M University System. He sees hope in the form of academic studies on agriculture, the incredible research that fourteen different departments are conducting on crops and nutrition, and new discoveries such as water-saving irrigation projects. These advanced technologies are being developed and put into practice in agriculture.

---

14 Joseph Hincks, "The World Is Headed for a Food Security Crisis. Here's How We Can Avert It," *TIME Magazine*, March 28, 2018.

15 "Farming and Farm Income," Economic Research Service, USDA, updated February 7, 2024, https://www.ers.usda.gov/data-products/ag-and-food-statistics-charting-the-essentials/farming-and-farm-income/.

16 Debbie Koenig, The Future of Food," WebMD special report, June 2, 2022, https://www.webmd.com/diet/future-of-food-special-report/20220602/the-future-of-food.

Let's remain optimistic, but we must not be naive or sit on our hands doing nothing. It will take all of us in our unique way—government, corporations, philanthropists, members of faith communities, nonprofits, and the public—to ensure food security for all. Despite the billions of pounds of food distributed through the charitable network, a considerable gap exists. There is much work left to be done.

Volunteers

Another vital group of people who donate to food banks are the volunteers who contribute their time. Once, a group of second graders volunteered to collect food for the food bank where I worked. They presented us with one of the most special volunteer donations ever: a brightly colored cardboard box decorated with pictures, glitter, and streamers. As I looked into the box, I saw that string beans barely covered the bottom. The students beamed with pride as they told me they grew these beans as a school project and wanted to feed the hungry. They wanted to donate their "first harvest" to "Second Harvest." Out of my decades in the nonprofit world, that is one story I will never forget.

Volunteers are essential to the success of a nonprofit. Every hour makes a big difference. I often heard people saying, "Oh, I'm only a volunteer." They undervalue their worth to the cause. In many nonprofit sectors, most organizations depend heavily on or are run by volunteers. So, if you question what one person can do, think of this African proverb, "If you think you're too small to make a difference, try sleeping with a mosquito." The collective power of volunteers is significant. According to Independent Sector, an hour of a volunteer's time is worth $31.80. Collectively, including all nonprofits' volunteer time, the annual value totals nearly $200 billion nationwide. [17] The Second Harvest Food Bank of Central Florida tracks their volunteer hours; in a recent year, that time was worth just over $3 million! That's $3 million saved from payroll.

---

17   "Value of Volunteer Time," Independent Sector, April 19, 2023, https://independentsector.org/resource/value-of-volunteer-time/.

Volunteers come in all shapes and sizes, but all desire to care for the health and welfare of people in their communities. One out of four Americans volunteers. They are men, women, retirees, teenagers, former recipients, professionals, corporations, members of faith-based groups or school groups, students, and dedicated folks with special needs. They volunteer for different reasons, such as sharpening skills to reenter the workforce, staying active during retirement, meeting new people, serving their communities, and living out their values. While volunteering, they learn new skills and meet new people. I have seen situations where people met while volunteering and got married ... at the food bank!

Volunteering broadens your perspective and can put your dilemmas and disappointments in perspective. It can boost your self-confidence, self-esteem, and life satisfaction. As I mentioned in the introduction, my life trajectory was changed because I volunteered decades ago at the food bank in Miami. I went from pursuing a career to embracing a vocation. Believe it or not, volunteering can impact your health. Volunteer activities keep people moving and thinking at the same time. Mayo Clinic Health System studies found that volunteering among adults aged sixty and over benefits physical and mental health.[18] Volunteers report feeling a sense of meaning by spending time involved in service, both given and received, which can have a stress-reducing effect.

I believe people realize a piece of their self-actualization when they volunteer. While folks are becoming the best they can be, they are also helping others meet some of their essential needs. Most volunteers desire to be part of something bigger than themselves and to derive a sense of meaning. In 2006, Al Delio started volunteering at the food bank in Orlando. Since that time, he has donated over 1,200 hours of his time. He is the volunteer captain of a team that sorts up to ten thousand pounds of donated meat daily. Al said in an

---

18   Angela Thoreson, "Helping People, Changing Lives: 3 Health Benefits of Volunteering," Mayo Clinic Health System, August 1, 2023, https://www.mayoclinichealthsystem.org/hometown-health/speaking-of-health/3-health-benefits-of-volunteering.

interview with WMFE radio that it is "ironic because I'm a vegetarian." Delio, a retired engineer, volunteers about thirty-five hours per week, which gives his life purpose. "After working 45 years, now I can volunteer 45 years, God willing, and I can give back."[19]

Volunteering is also beneficial to the young. I remember years ago when my grade school–age son volunteered to ride along with one of my food bank drivers and deliver food to homeless shelters and drug rehab centers in the inner city of Miami. I could have told him who needs help and what it's like in the real world for some people, but that would have been nothing like seeing it firsthand. For him, it was an eye-opener because we lived in a lovely community, and he attended a private school with friends who had wealthy parents and privilege. It made him realize that not everyone lived like our family. There's nothing like broadening perspective through life experiences. Parents play an essential role; when the kids see Mom and Dad contributing, there's a good chance the children will model that behavior as they become adults.

Volunteers seek credibility in the organizations they work for; they don't want to waste their time on nonperforming organizations. They also want an enjoyable and productive experience. Recognition of their efforts and communicating the impact they're making is essential in ensuring that they will continue to volunteer in the future. Mindy Ortiz, the director of Volunteer Services at the food bank in Orlando, said, "There's something extraordinary about a group that has gathered to build kids' food packs specifically. The energy in the room is different. When I get in front of a group and share how the Kid's Pack program came to be and talk about children's bellies hurting and their inability to concentrate in school because they have nothing to eat, the volunteers can relate. It is 'game on' to build the kids' packs." Volunteers enable nonprofit organizations to keep their doors open in

19   Danielle Prieur, "Second Harvest Volunteer Al Delio Says Feeding Struggling Families During the Pandemic Has Given Him New Purpose," WMFE interview, December 21, 2021, https://www.wmfe.org/2021-12-21/second-harvest-volunteer-al-delio-says-hes-driven-by-a-simple-bible-verse-whomsoever-much-as-given-much-will-be-required.

many instances and allow the organization to deliver vital programs and services.

An essential class of volunteers that are greatly appreciated are the folks donating their professional skills, such as board and committee members. A nonprofit needs every service a business does, like financial expertise, lawyers, and CEOs. The food bank benefits from food service professionals, commercial real estate talent (when finding property for a new site), logistics professionals, subject matter experts, and people knowledgeable in finance, law, education, and more. Over thirty million men and women in this country serve on the board of a nonprofit. Board members are critical in overseeing the health of an organization. The relationship between the board and the organization's leader is one of the most critical aspects. Over the years, many of my colleagues and nonprofit directors have shared how dysfunctional their board relationship is sometimes. This dysfunction often leads to the organization's underperformance and an exodus of talent from both sides. It must be a shared leadership situation. I liken it to a dance; sometimes you lead, other times you follow. Don't step on each other's toes. Find the rhythm.

Corporate groups often donate professional services. One great example is some folks who came to sort food from Lockheed Martin, a prominent local and national employer. During their break, a conversation with staff led to them discovering how they could help even further. One of the volunteers was responsible for teaching Six Sigma at Lockheed. The training is all about process improvement and measurement of impact. This kind of training is typically unaffordable for a nonprofit, so Lockheed voluntarily trained thirty-five food bank staff members from various departments. This has resulted in professional development for the staff and greater efficiency of the food bank, all at no cost. In another example, Disney provided help from Imagineers to create incredible murals throughout the food bank, and its industrial engineers conducted time studies to enhance our operations in the distribution center. Universal Orlando provided professional chefs and research services, and UPS carried out route optimization for our fleet of trucks, saving the food bank thousands of gallons of fuel. I can think of numerous other examples. I refer to it as an "embarrassment of riches." Many markets don't have access to a major entertainment complex or a Fortune 500 company, however, professional experience resides in some form in almost any city or town.

Education
Institutions of higher education are also great partners. The University of Central Florida staff have served on the Second Harvest Food Bank of Central Florida board of directors. The school provided critical hunger research in the local community that helped us become more knowledgeable and, as a result, more strategic in our approach. Valencia College in Orlando offers credits to our graduating culinary students should they want to continue their education. Rollins College, a premier, private liberal arts college in Winter Park, Florida, continues to be a highly valued partner with the Second Harvest Food Bank of Central Florida and have chaired the board. The Rollins MBA students have dedicated semesters to projects for the food bank. I once asked if a class could calculate the annual economic impact of the food bank. They embarked on a deep study, analyzed all the functions of the food bank, and presented a detailed report. Their finding: The food bank had an estimated economic impact of $187,389,896 in 2016. And that number has grown since the study was completed in 2016.[20] It is based on metrics that include the environmental impact of products diverted from landfills, tons of CO_2 avoided, and healthier ambient air. Other factors were the value of tax incentives for food donors, the in-kind value of the food, the operating budget, and the various programs, volunteer time, and SNAP benefits generated. The collective economic impact of the two-hundred-member Feeding America network must be in the billions of dollars. Food banks across the country that are located in college

---

20   Keenan Yoho, PhD advisor, "Second HarSecond Harvest Food Bank of Central Florida Economic Impact Study", MBA project, Rollins Crummer Graduate School of Business, 2016.

towns can tap into the resources of these schools. Colleges are looking for ways to help their communities and engage their students in meaningful experiences.

The Media

One of the most valuable in-kind "donors" is the media, whether television, radio, or print. Ideally, being able to benefit from all of these is incredibly powerful. This in-kind coverage does not consist of paid media but stories about the cause and the nonprofit. From personal experience over the years, I can say that food banks have a great opportunity when relationships are built with media partners. Most importantly, though, the nonprofit must have a newsworthy story. There are tens of thousands of nonprofit organizations across the country, and hundreds within a community; however, most people cannot name more than a handful. How can a nonprofit gain support without visibility? It cannot. The public wants "good news" stories. Without awareness, interest cannot be generated, the desire to help is lacking, and action does not happen.

Through media relationships, there is an opportunity to become a recognized nonprofit. Various TV and radio stations can post frequent stories regarding the food bank's innovative programs and conduct interviews with people receiving food. Newspapers offer similar possibilities, as well as opportunities for nonprofit staff to provide their own opinion pieces. Over time, these mentions add up to a consistent awareness level. In the end, thousands of our neighbors facing hunger are helped.

At the national level, Feeding America connects with major media partners to communicate about the hunger issue, and this coverage links a substantial part of the solution to food banks. A story on any of the major networks reaches millions of people. Recall that during COVID-19 or after hurricanes, floods, and tornadoes, there are always images of thousands of cars lined up to receive food. People respond to stories like this by offering more help. Like many others, the Second Harvest Food Bank of Central Florida is fortunate to have powerful partners. One of many examples is our local Central Florida NBC affiliate, WESH 2, and its powerful champion for hunger relief, Executive Producer Marcie Golgoski. For more than three decades, the station has devoted an entire week of programming each year to highlighting the issue of food insecurity and providing an easy way for people to contribute toward a solution. The Share Your Christmas food and fund drive takes place at the time of year when people are thinking about giving and has provided more than $2 million over the years. Leading up to that point, the food bank spent years cultivating that relationship and providing valuable information to the station on the hunger issue. Twice during COVID-19, WESH 2 held an on-air marathon requesting donations for the food bank because of the unprecedented need. Well over $1 million came in.

Celebrities

Another way to leverage media coverage of a nonprofit is through the support of actors, sports stars, and entertainers, should they be present in your area. Their renown and popularity can shine a light on a worthy cause. Feeding America has an entertainment council that supports hunger relief with names such as Rachael Ray, Matt Damon, Sheryl Crow, and Ben Affleck. Scarlett Johansson states, "My family grew up relying on public assistance to help provide meals for our family. Child hunger in America is a real and often overlooked problem, but one together we can fix. I hope that by joining Feeding America, our awareness of this issue will lead to a solution."[21]

Other stars help bring attention to the issue of hunger. An example of a profound commitment came from Harry Chapin, a singer-songwriter famous in the 1970s. Before his untimely death, Chapin became involved in the fight to end hunger. He cofounded WhyHunger and established a number of food banks. After his death in 1981, he was awarded the Congressional Gold Medal for his food advocacy work. Another example of using celebrities to highlight the cause of hunger is found in John Lennon's song "Imagine," which speaks of a utopia without issues, including hunger. Yoko Ono worked with

21   "Actor Scarlett Johansson," Feeding America, November 8, 2023, https://www.feedingamerica.org/partners/entertainment-council/scarlett-johansson.

Feeding Florida to design a signature license plate based on John's work, with proceeds supporting hunger relief.[22] More recently, Taylor Swift made substantial donations to food banks in the communities she visited on her 2023 tour. Her donations help in two ways: by augmenting the food bank supplies and attracting wider attention to the problem of hunger.[23]

My all-time favorite story of a celebrity supporting hunger relief is the Boss, Bruce Springsteen. Over the years, starting with the Community Food Bank of New Jersey, he has donated part of the proceeds of his concerts to hunger relief. When I led the food bank in Miami in the 1990s, he came to town. His agents informed us that we would receive a donation and an opportunity to see him live in concert and backstage after the show. I am a huge fan of Bruce; this was quite the moment. After the concert, two food bank staff members and I went to the green room where celebrities and friends of Bruce congregated to see him. It was packed wall to wall, and my hopes dimmed of ever getting to him. Suddenly, his agent asked for the "food bankers"; Bruce wanted to see us first. We went to a small back room, sat at a table for four, and anxiously waited. Bruce walked in with a bottle and four glasses, plopped them on the table, and asked, "How are you doing?" We spent fifteen minutes with him; I cherish my photo of us as best buds. His songs tell the stories of blue-collar folks, immigrants, and social issues; it was unique, to say the least, to see that come out so naturally in the conversation. The "bromance" continues.

I'll end this topic with a heartfelt "thank you" to volunteers; nonprofits need you and appreciate you. Consider the following note, which the food bank received from a food recipient, as a tribute to volunteers:

I saw your food pantry today and was given some groceries. This was the first time in my life that I requested any aid from any organization, and I would never have

---

22   "Imagine Fighting Hunger," Feeding Florida, 2023, https://www.feedingflorida.org/taking-action/imagine-tag.

23   Tami Luhby, "Taylor Swift is a Hero to Food Banks from Coast to Coast," CNN, August 13, 2023, https://edition.cnn.com/2023/08/13/business/taylor-swift-food-bank-donations/index.html.

approached your organization if it wasn't essential. I lost a job, have very little savings, and have had no luck finding a new position for over two months. Despite my best efforts, I'm down to my last thirty dollars. Anyway, when I got home, I studied all the food on my dining room table. I was astounded. I realized that some unknown person, a volunteer, had made a genuine effort to assemble a sustaining and logical package of food from all the random donations you must receive. It wasn't just a lucky coincidence.

In the past, at school and then at work, I've always donated to food drives. Okay, a can or two of this, a box of that, and hopefully, someone will eventually get some good out of it. But I was very casual about it, thoughtless. It was almost abstract. I just never realized how meaningful a can of lentil soup might be.

Anyway, I want to thank your organization for this beautiful array of food on my dining room table. I wish you could see it! And most notably, I'm thanking that unknown person who cared enough to think about putting a can of tomato sauce in with the bundle of pasta, applesauce for fruit, and even cookies. It is meaningful, and your effort is understood and so appreciated.

Ethics

Nonprofit organizations can run into ethical and moral dilemmas when working with donors. Consider these examples. What would you do?

Gallagher, a comedian, used to destroy watermelons on stage and splatter the first few rows of the audience (they were given ponchos). Would you accept part of the proceeds from his show? Would a few ruined watermelons be worth a donation of a few thousand dollars? Or do you stand on the principle that wasting food is not correct? Would you accept a donation from a payday lender that charges exorbitant fees to its low-income customers? What if the Proud Boys wanted to help and expected high awareness in exchange for their support? What if a vodka distillery is launching a new brand to high-end customers and wants to donate part of the proceeds from the sales to your organization? Remember that some places you distribute food to are nonprofit substance abuse facilities and homeless shelters. Do you accept those dollars? Would you do it for $1,000, $2,500? $10,000? What's your price? Do you have a rationale for whatever decision you make?

I once heard that a church returned a $1 million donation when they found out it was from the owner of a casino. Another time, Slap My Ass and Call Me Santa, a hot sauce company, wanted to donate the proceeds of the sale of this product to the food bank and tie in a high-visibility promotion. After considering that the audience for the product was adults and it would have very limited or no exposure to kids, I accepted the donation and fed more people. It's important to be proactive and prepare for these decisions as a leader; decide where you draw the line on ethical decision-making. Have some solid guidelines or moral code that you and the board have established.

Ultimately, various contributions make food banks successful. Funds from donors provide vital operating expenses, while farmers, retailers, manufacturers, and the federal government donate vast quantities of unused food each year. Other people give the food banks what they can in the form of time and talent—saving on expenses for the organization and weaving a wide network of people invested in successfully addressing hunger in the United States.

Part 3

How Are People Being Helped?

Chapter 9
Innovation

*Have you ever seen innovation where
someone didn't take a risk?*
- Jean Case

There's not one definition of innovation. It revolves around creating something new of value or merging the old with the new. It means addressing a challenge or opportunity and imagining possibilities. It comes from a deep-seated belief that there's got to be a better way. And there are different types of innovation. One is incremental innovation; many small positive steps can lead to significant results. Tweaks in operational processes lead to constant improvement, time and money savings, and better quality of the service or product. The other type is groundbreaking innovation: reframing something by asking the big "why" or "why not" questions. This type can lead to systemic change. Former president Jimmy Carter said, "Go out on a limb; that's where the fruit is."

Innovation Abounds
One of the many strengths of the Feeding America national network and its two hundred members is its innovative spirit. Innovation is essential because resources are always limited, so it is necessary to become creative in many ways to stretch the dollar. Innovation has occurred in food banks over the forty-plus-year history of the Feeding America national network. Food banks started in the early 1980s, collecting good food

that would otherwise be thrown out and giving it to people in need. That initial concept was innovative then: food in, food out. Reduce waste, feed people. Fast forward to today—Feeding America food banks are running countless programs, connecting people with public benefits like SNAP and offering various types of training (truck driving, culinary, warehouse vocational skills, etc.). They are running job placement programs and mobile pantries, carrying out online ordering for partners and neighbors, operating social entrepreneurship ventures, and more. They are feeding people today and preparing them to feed themselves tomorrow. Contemporary food banks barely resemble what they were decades ago. They will always be about food distribution, but they are addressing it in many new ways. One of the great strengths of the Feeding America network members is their ability to question the status quo. Challenging times demand new solutions. We all have a choice in tough situations: do we step back and accept the status quo, or do we step forward and do something about it?

Innovation is vital to the effectiveness of any organization. Solutions to vexing social issues can only be fixed with fresh thinking and action. It bears repeating that innovation means creating something new and valuable; it is transformational and forward-thinking. Innovation is not about making an already intolerable situation marginally better. Instead, it is about working with our stakeholders on the issue, yielding designs that feel like pieces of an entirely new, emerging world. The status quo has much power and is risk averse. Innovation is hard work because it means swimming against the current of a dominant culture. The inclination is to accept things at face value and not question what is behind the thought or rationale.

Source and Environment Promote Innovation

So, what conditions are needed to promote innovation in an organization? The Second Harvest Food Bank of Central Florida's culture statement includes a declaration that says, "We innovate today to create our tomorrow." It is best to create the right environment where anything can grow. The conditions transform the people and systems involved, expanding their creative capacity for future innovation. In other words, for innovation to be successful and viable within an organization, it must be in its DNA, embedded deeply into how the organization thinks, acts, and behaves. Permission to fail is one of the most important criteria. Brené Brown, author, professor, lecturer, and podcast host, describes the conditions that inhibit innovation: "No vulnerability, no creativity. No tolerance for failure, no innovation. If you're not willing to fail, you can't innovate. If you're not willing to build a vulnerable culture, you can't create."[1] Innovation will come with mistakes at times; we all have scars. But that doesn't mean we've failed; we made it through the battle.

Innovation can and should come from any level of the organization—ideally from the bottom up, because we must respect the experience and knowledge of staff members, colleagues, customers, and clients. It is a liability to have only one perspective. Diverse perspectives—including solid representation of voices often excluded or silenced—are needed to generate innovative insights. A great example of getting the proper perspective of others comes from a national task force I co-chaired with Feeding America. The initiative was called Reimagining the Neighbor Experience. Neighbors, as defined by Feeding America, are people in challenging circumstances who need food relief. For food banks, this initiative was akin to transforming the charitable food system. For decades after food banking started, the primary "customers" were partner agencies such as pantries, shelters, and soup kitchens. Those partners interacted directly with the people in need. For years, food banks paid little attention to their role with the neighbors. Food banks learned that their partner agencies were under-resourced, and most could not deepen their relationship with those served. The food banks wanted to know how to come alongside the partners and complement their great work. Feeding America's approach to this effort was rigorous; they combined innovation and research expertise to address the challenge. The critical question was, what does putting the

1   Brené Brown, "Brené Brown, Vulnerability, and Strong Leadership," Leading Edge, April 25, 2019, https://www.leading-edge.org/post/brene-brown-vulnerability-and-strong-leadership.

neighbor at the center of our work mean? Through extensive conversations with our neighbors, so much came to light that is being used to design future programs. One example is that food banks are now using technology and systems like Grubhub and DoorDash to increase access, convenience, and dignity when providing food. This eliminates the need for a person to visit a food pantry. It also allows outreach to people who would never visit a pantry due to social stigma.

Innovation can also come from co-creation, including with outside partners. It can happen at the micro level of our neighbors, the interdepartmental level of organizations, or the community partner level. Optimally, it happens at all levels. For example, at the community level, the Second Harvest Food Bank of Central Florida co-created a program with a significant donor, Universal Orlando. During a lunch I had years ago with the VP of Community and Diversity Relations at Universal Orlando, she said they wanted to create a signature program that would live on and engage their employees in a big way across the company. We conducted a brainstorming session with Universal's senior executives, who oversaw areas such as food service, entertainment, research, human resources, diversity, creativity, and operations. The food bank and Universal were focused on getting food to middle school students beyond their school lunches. This was an underserved population, and their egos were too big to accept a backpack of food; they were too cool. The solution was to design a mobile pantry that looked like a hip food truck: brightly painted, as well as with music. Universal chefs accompanied the truck and passed out food samples while volunteers provided food bags. Food demonstrations, menus, and nutritional education took place in some classrooms, led by chefs. As a result of that co-creation, the program has gone beyond delivering its goals and lives on today. We continued to learn from the program rollout and tweak the model as needed. This example is a rallying cry to donors to be open to co-creation and nonprofits. It all translates into a more significant community.

The food bank has also engaged in co-creation with the Morgridge Family Foundation. This foundation invests in leaders and organizations that are reimagining solutions to today's biggest challenges. It places the highest value on innovation and collaboration and has a long, proven track record of success. It brings different sectors to the table, such as business, government, education, and private funding, to foster innovation as a vehicle for systemic change. One of the foundation's new initiatives, currently in the planning stages, involves looking at the traditional way of providing food for our neighbors and turning it on its head, genuinely pushing the envelope. If successful, this initiative could somewhat reduce the demand for food banks.

The project aims to create a food utility that would provide free food for all, and it is in the early stages of conceptualizing a new food ecosystem. A customer would walk into a grocery store and choose from one hundred preselected items that would be free. Stigma would be removed from accepting free food, while the dignity of the shopping experience would be enhanced. The foundation is doing its due diligence in talking with the local communities where the program may be piloted to ensure the types of food locals want are included. They are also engaging nutritionists, local growers, and grocery retailers. This concept is undoubtedly pushing boundaries and has enormous obstacles to overcome. However, no matter where the effort ends, there will be discoveries, new partnerships, and a better understanding of the charitable food supply chain. The end product could be a hybrid solution that combines the original concept with newly discovered elements.

Positive Turbulence

Innovation can also be borne of necessity. I've been a member of the Association of Managers of Innovation (AMI), an incredible group of people worldwide who focus on innovation in their careers. The AMI's inspiring leader and founder, Stan Gryskiewicz, crafted an approach to solving problems in his book titled, "Positive Turbulence."[2] Rather than throwing up

---

2   Positive Turbulence, copyright 1999, by Stanley S. Gryskiewicz, Jossey-Bass Inc.

our hands in desperation and waiting for someone else to step up and solve problems, we can use challenging circumstances and situations as opportunities to respond creatively. Positive turbulence recognizes that change is inevitable and seeks novelty from the outside. A changing economy, new technology, unknown competition, shifting population demographics, environmental disasters, partisan politics, and social trends surround us. Half the battle of solving complex issues is just showing up with the right attitude. Mary McBride, an AMI colleague of mine, offered this sage wisdom during an interview for the book: "The energy of creating a vision is constantly informing us, so we shouldn't shut it down and only think about what's within our logical and literal minds. It's more than intuition, but the humility to understand that we're not alone in creating and sit with that fact for a moment and let's see what energies show up to guide us."

Disaster Relief
Sometimes the energies that guide are enormous. I had never seen or personally experienced anything like Hurricane Andrew. The storm initially hit the Bahamas. On August 24, 1992, it landed in South Florida and then traveled north to Louisiana, causing unprecedented damage, $27 billion in losses, and at least sixty-five fatalities.[3] The evening that it blew in was unforgettable. I gathered my wife and two young children, and we stayed in the most secure room of our house, one of the bathrooms, for hours. The wind sounded like a freight train, electrical transformers were exploding, trees were falling, and objects were hitting our house. I wasn't sure if the plywood over the windows would hold or what I might find outside in the morning. Fortunately, we only experienced an extended power outage, blocked streets, and tree damage in our area, forty miles from the eye of the storm.

Others were not so lucky. Hundreds of thousands became homeless. Housing developments were so flattened that they were unrecognizable. Some residents in the flooded areas

---

3   "Hurricane Andrew," National Weather Service, National Oceanic and Atmospheric Administration, 2022https://www.weather.gov/lch/andrew.

who still had part of their homes left had fish pouring out of their kitchen cabinets. I remember mobile homes being wrapped around telephone poles like tinfoil, and a palm tree that was speared with a blunt wooden two-by-four. An older woman thought she heard a looter in her garage after the storm and grabbed her shotgun. As she flung open the door to the garage, she saw an escaped baboon from the Miami Zoo, one of the hundreds of wild animals on the loose. It was a cleanup of nightmarish dimensions, and it took more than a year for homes to be rebuilt, roofs replaced, and an entire area's infrastructure restored.

Not long into the relief effort, I became the interim executive director of Daily Bread Food Bank in Miami, a baptism by fire. I was in Homestead, Florida, ground zero, alongside a new colleague, Mike Mulqueen (CEO of the Chicago Food Depository), who had flown in to help out. Mike provided his identification as he and I approached the guarded gate to a military compound to get help with relief efforts. The guard promptly snapped his heels together and saluted. I had no idea that Mike was a retired four-star general in the Marines! He was precisely the person you wanted on your team. We needed help distributing food, in the form of trucks and labor; a few days later, military trucks and soldiers arrived at our food bank loading docks, ready for action. Peacekeeping was one of the most significant benefits of having a military presence. Without them, I can't imagine the danger and destruction that would have occurred.

The aftermath of the hurricane was full of both difficult and wonderful moments. One of the most unexpected but touching experiences occurred in December 1992, just a few months after Hurricane Andrew's landfall, amid significant relief efforts. Police escorted my family and a dear friend to a migrant camp one evening to distribute food and gifts to the families that were so hard hit. Because of winds and flooding, these farmworkers had no crops to harvest and, as a result, no income. The camp was dark due to the lack of streetlights and loss of power. As my young son frantically looked for one more toy for a waiting child in the back of our minivan, I heard what sounded like angels singing. As I looked toward where the sound was coming from, a procession of twenty to thirty children with candles came around a corner, singing Christmas songs in Spanish. I realized I was hearing angels sing!

During the recovery period, Feeding America members worked alongside the Salvation Army and the American Red Cross for months, sharing a 250,000-square-foot distribution center in South Florida. Hundreds of millions of pounds of food flowed through that facility. Much of the help offered was from individuals. The flip side of the chaos and destruction was a robust phenomenon: strangers' kindness. There are countless stories of people who performed selfless and heroic deeds over several months to help their neighbors. All the while, they had experienced damaged homes and lost jobs. However, we received some other help that was less successful. For example, we received a truckload of old fur coats in Miami for hurricane relief and enough water donations to create major artificial lakes.

Until Hurricane Andrew, food banks didn't see themselves as disaster relief organizations. However, they have proven to be a critical factor in successful relief efforts ever since. The Feeding America national network now has a plan for responding to all types of disasters, whether they are local, regional, or national in scope. One innovation is the pre-positioning of supplies in disaster-prone areas so that relief can be quicker. Additional measures include the creation of infrastructure for refrigeration and emergency power generators, local disaster relief plans and exercises, setup of drive-through food distributions at a massive scale, home delivery of meals, and much more.

Many other disasters, including fires, floods, freezes, earthquakes, hurricanes, and tornadoes, followed Hurricane Andrew. Hurricane Katrina hit the Gulf Coast on August 23, 2005, as a category-five storm. It hit hard, particularly in New Orleans and surrounding areas, causing about 1,200 fatalities and an estimated $75 billion in damage.[4] The food banks

4   "Hurricanes in History," National Hurricane Center and Central Pacific Hurricane Center, National Oceanic and Atmospheric Administration, 2022, https://www.nhc.noaa.gov/outreach/history/.

contributed their newfound expertise for months after that devastation and learned additional lessons. Brian Greene was the executive director of the New Orleans food bank then; he and his staff worked tirelessly for months, along with food bankers from across the country. At this time, the food bank network mainly worked behind the scenes during disasters without much visibility; now, they have become a destination for volunteers and a channel for the goodwill of America. According to Brian, "it's almost limitless what a food bank can do. At a moment's notice, they are organizations that distribute two to three times the average volume of food; we're talking about tens of millions of pounds. They take a leap of faith in getting resources." Over the years, the Feeding America members have developed working relationships with their local and state emergency management offices, FEMA, the National Guard, and organizations such as the Salvation Army, Red Cross, and the United Methodist Committee on Relief. And behind the food bank network is the immense food industry that supplies billions of pounds of food and water during disaster relief. Individuals, faith-based groups, and the business sector across America and internationally generously donated financial resources to make relief efforts possible. The Feeding America network combined their individual strength as members to produce a powerful, collective response. Better together.

COVID-19

Another example of turbulence promoting innovation arose with the outbreak of COVID-19. This unprecedented event required food banks to change their distribution model on the fly. Food banks are set up for large-scale food distribution, ranging from four million pounds for a rural food bank to just over one hundred million pounds in a major metropolitan area. That requires distribution at food pantries that serve hundreds or thousands of people. When COVID-19 hit, people were self-quarantining; many were afraid to go to the grocery store and be infected; others didn't have transportation. Food banks started to become overwhelmed with the volume of calls and requests for food, many from people who had not eaten in a few days. Many of their pantries were shut down. The remaining distribution network of pantries could not meet the demand because of the high volume and sudden loss of volunteers. I recall speaking to one of my team members who was answering calls and hearing the heartbreaking stories of desperation combined with embarrassment about needing help. Many of these callers had never needed to ask for service before. Most of the calls were from seniors who could not afford food, did not have transportation, or had a physical or mental disability.

As a food bank, we could only reach these people if we developed a new solution. The idea of home delivery surfaced; some staff balked because we had never done that. The sheer volume of thousands of calls for food from people who had not eaten in days was the final impetus to try this solution. If it failed, we would learn quickly and adapt. Sometimes, you must step out with courage and confidence, and eventually, the resources will appear. You cannot wait until you know everything.

Our initial launch changed quickly as we gathered more information from the people we served. We prototyped early and often. In the end, we created a program (Bring Hope Home) that responded to requests in a vetted, timely manner and delivered food directly to people's doors at no cost to them. More than a year since its launch, Bring Hope Home has distributed over one million home-delivered meals and become a permanent program. We listened to the outside, the periphery, and the neighbors in need. It was a program I could not have imagined before COVID-19. Necessity is the mother of invention.

Health and Hunger
Many disasters do not descend as quickly as hurricanes or pandemics, but are equally deadly to the population. In one major example of innovation, food banks have sought ways to address obesity. This country is home to food insecurity and its equally evil twin, nutritional insecurity. On the one hand, too many people lack food access and the income to sustain their most basic needs. On the other hand, many unhealthy, primarily

cheap, and sometimes addictive high-calorie and high-carb foods are the most affordable and accessible. According to the Healthy Food Playbook, lack of access to healthy food lessens people's ability to eat well. In addition, it increases the rates of obesity, diabetes, and other diet-related conditions.[5]

Rates of obesity are increasing. Levels were fairly static during the 1960s and 1970s but then rose dramatically, from 13.4 percent in 1980 to 41.9 percent in 2020.[6] Obese children are likely to become obese adults, as they exercise less and spend more time in front of screens. Obesity is a major factor in many illnesses, like hypertension, cardiovascular disease, and, most notably, diabetes, which is the seventh leading cause of death and costs billions of dollars in medical costs and lost wages. People diagnosed with diabetes have more than twice the average medical costs of those without. The Centers for Disease Control and Prevention reports that 37.3 million Americans have diabetes. About one in five people with diabetes don't even know they have it.[7] Obesity is rarely random. In 2022, the Trust for America's Health reported that social and structural factors in American society largely determine the national rates of obesity. Poverty, housing insecurity, discrimination, and lack of access to high-quality food and health care are all significant drivers in creating an obese population.[8]

Other studies make it evident that economic insecurity affects health beyond obesity. In 2014, a report linked food insecurity to hospital admissions.[9] The study correlated income

5   "Delivering Community Benefit: Healthy Food Playbook," Health Care Without Harm, communitycommons.org/entities/o57cef/1-0508-451a-975b-3d4f588b0217

6   Office of the Surgeon General, "Background on Obesity," *The Surgeon General's Vision for a Healthy and Fit Nation*, 2010, https://www.ncbi.nlm.nih.gov/books/NBK44656/; "Adult Obesity Facts," Centers for Disease Control and Prevention, https://www.cdc.gov/obesity/data/adult.html.

7   "Diabetes Basics," Centers for Disease Control and Prevention, last modified May 17, 2022, https://www.cdc.gov/diabetes/basics/.

8   "State of Obesity 2022: Better Policies for a Healthier America," Trust for America's Health, September 27, 2022, https://www.tfah.org/report-details/state-of-obesity-2022/.

9   Hilary K. Seligman et al., "Exhaustion of Food Budgets at Month's End and Hospital Admissions for Hypoglycemia" (abstract), National Institute of Health Public Access, 2014, https://www.ncbi.nlm.nih.gov/pmc/articles/PMC4215698/pdf/nihms575875.pdf.

to hospital admissions for hypoglycemia. Hypoglycemia—a common side effect of lack of food and diabetes—can cause acute symptoms and traumatic accidents. In examining the nearly twenty-five million patients admitted to California hospitals over eight years, the study discovered that "admissions for hypoglycemia were more common in the low-income than the high-income population."[10] Even more telling was that the timing of admissions seemed related to food access. It was stated in the report that "food budgets are exhausted at the end of the month than at any other time ... The risk [for admission] increased 27 percent in the last week of the month. SNAP benefits are typically distributed in the first ten days of the month." Finally, among the poor, hospital admissions were associated with the cycle of monthly expenditures and when they ran out of money: "Most fixed expenditures are also paid out in the early weeks of the month, leaving less money available at the end of the month. Approximately half of all households making mortgage payments or rent do so between the last day of the month and the seventh day of the next month."[11] The report concluded that the exhaustion of food budgets was a factor in health inequities that could be addressed through different food and benefit access policies, as well as greater national emphasis on food insecurity. As one author, Hilary Seligman, the senior medical advisor to Feeding America, mentioned to me during our interview: "This is a solvable problem; we know the answers; funders need to unite with the government, private sector, and the faith community. If we put the same focus and resources on hunger as on COVID, the results would be transformational. Diet-related diseases will cause significant long-term damage to this country."

Part of the damage done by these hunger-induced illnesses is financial. In 2011, hunger cost the United States at least $167.5 billion in a combination of charity, avoidable healthcare costs, and lost economic productivity due to work

10   Seligman et al., "Exhaustion of Food Budgets" (abstract).
11   Seligman et al., "Exhaustion of Food Budgets at Month's End and Hospital Admissions for Hypoglycemia," Health Affairs, 33, no. 1 (2014):116-123

absences.[12] Providing healthy food to someone for a few hundred dollars a month to avoid a hospital stay certainly makes sense, considering that the average cost of an emergency room visit is $2,600.[13] This cost does not include the billions spent annually to care for patients who are readmitted to the hospital within thirty days for a previously treated condition. As recently as 2018, the average readmission cost was $15,200 per patient.[14]

Hunger and dietary illnesses also threaten our nation's security. During the conflicts in Afghanistan and Iraq, General Mark Hertling was assigned the responsibility of improving basic training and reducing the high attrition rates. This attrition was mainly due to the soldiers' poor physical condition. During our interview, he stated that 75 percent of civilians were not qualified to enter the military, primarily because of obesity. Of the 25 percent who could qualify, 60 percent could not pass the physical training test on the first day—one minute of push-ups, one minute of sit-ups, and a one-mile run. Those who did join the Army were used to a diet of fast food and soda, as well as spending up to sixty hours per week in front of screens. This lack of fitness resulted in high levels of injuries among recruits. Many had poor bone density and were prone to trauma injuries like stress fractures and pelvic cracks. The Army spends between $100,000 and $300,000 on average medical costs for each operation.

Several changes were implemented, which yielded positive results within a year. Hertling changed the culture to one focused on creating soldier-athletes. Dining facilities were revamped, and deep-fat fryers were removed. Danishes and chocolate desserts were replaced with fruits and veggies,

---

12   Donald S. Shepard, Elizabeth Setren, and Donna Cooper, "Hunger in America: Suffering We All Pay For," Center for American Progress, October 2011, https://www.americanprogress.org/article/hunger-in-america/.

13   "ER, Urgent Care or Virtual Visit? What to Consider to Help You Save on Costs," United Health Care, March 2023, https://www.uhc.com/news-articles/benefits-and-coverage/care-options.

14   Caroline Fallon, "4 Reasons Why a Hospital Readmission Rate Matters," Comprehensive Medication Management, March 2023, https://blog.cureatr.com/4-reasons-why-a-hospital-readmission-rate-matters.

and soda with water. The general said it was tough telling a soldier to eat salad! Year one resulted in reduced injuries, saved $30 million in treatment, and resulted in healthy weight losses. Hertling believes that food security and lack of proper nutrition is a readiness issue for the military: a national security issue.

Feeding America food banks started exploring the intersection of health and hunger around the mid-2000s. Before that, most food banks were not intentionally focused on nutrition beyond what they traditionally received as food donations. The bulk of the food consisted of shelf-stable packaged goods. When I became CEO of the Second Harvest Food Bank of Central Florida in 2004, I recall having a huge opportunity because we were surrounded by fresh produce. At one point, I spoke to an agricultural consultant and asked how much fresh produce in Florida is wasted yearly. His answer was 975 million pounds! According to a report by the World Wildlife Fund and the UK grocery chain Tesco in 2021, "farm-stage food waste is significant but often overlooked food waste hotspot."[15] The reasons for the waste include consumer demand for perfect-looking produce, last-minute cancellations of wholesale orders, lack of harvest labor, and crops that often do not thrive in a given environment.[16] Similar concerns led to food waste in Central Florida, with customers rejecting potatoes the size of footballs or carrots that looked like malformed fingers. At that point, the staff and board of directors took a closer look at fresh foods. I wanted to make a more significant impact in our geographic area.

Utilizing fresh produce was challenging. Food banks were initially equipped primarily for nonperishable foods and lacked refrigeration or freezing capacity. Also, a different base of expertise is needed to handle the many varieties of fresh produce, equipment, refrigerated trucks, and so on. As we planned for a new food bank facility, we designed it

15   World Wildlife Fund and Tesco, *Driven to Waste: The Global Impact of Food Loss and Waste on Farms*, 2021, 20, https://www.worldwildlife.org/publications/driven-to-waste-the-global-impact-of-food-loss-and-waste-on-farms.

16   World Wildlife Fund and Tesco, *Driven to Waste*, 18.

to maximize the number of fresh foods we could provide to people in need. We built massive coolers, hired professional staff from the produce industry, and equipped our partner feeding programs with refrigeration. I remember walking through the new facility before we transferred any food to it and looking at my COO as we peered into what seemed to be large coolers that were eighty by forty feet in size with thirty-seven-foot-high ceilings and saying, "We may have overdone our vision." We had to laugh about that moment when we were at maximum capacity a few months later.

Another challenge was conceptual. To what extent would we try to dictate the types of food our clients received? This discussion also happens nationally when the federal government assesses whether to limit SNAP recipients to certain foods based on their nutritional value. I asked Dr. Hilary Seligman her thoughts. She said:

> First, many restrictions exist on what people can receive through SNAP. From a public health perspective, there is no question that SNAP restrictions would improve health outcomes. The question in my mind is not can we do it? But should we do it? We have allowed high-calorie, nutritionally empty foods to be sold. They are marketed to kids incessantly. Poor food is a problem for all of us. The problem is not a problem for poor people; if we keep going for the easy fixes, there will be unintended consequences and further stigmatization. If you restrict for poor people, why not everyone? The cost is impacting all of us.

Eric Bost, the former undersecretary for Food, Nutrition, and Consumer Services at the USDA, also opposed banning certain foods in SNAP, although from a different perspective than Seligman. He argued, "It won't happen. I had Congress come to me and ask about banning certain foods; why don't we do that? I responded, 'Are you going to talk to the senator from Pennsylvania with Hershey's located in that state? Will you talk to the senators in Minnesota and Nebraska who grow our wheat? Will you talk to the senators from Idaho,

Georgia, and others about their cash crops not being included? So many foods include sugar, nuts, milk, et cetera. And don't lose sight that farmers are one of this country's largest, most reliable voting blocks."

To determine the depth of the food bank's commitment to focusing on health and hunger and make this decision for the food bank, I once drew a long horizontal line representing a continuum during a pivotal board meeting. One end of the line represented refusing to take sugared sodas, energy drinks, chocolate, candy, birthday cakes, and snacks. The other end meant taking all types of donated food, regardless of its nutritional value. I asked each board member to place a mark on the continuum of where they thought we should be. Most people voted in the middle, but the discussion included passionate voices from each end of the continuum. Some questioned how we could, with all our knowledge about proper nutrition and the diseases and deaths that occur due to poor food consumption, distribute unhealthy foods. From the other perspective, there were voices saying, "Who do we think we are, the judges of what hungry people eat? We don't even follow our self-professed diets most of the time." One very wise board member said, "We can't legislate if we don't educate."

The meeting inspired us to adopt a nutrition policy for the first time in the food bank's history. Our approach did not exclude all unhealthy foods but designated them as "sometimes food." For example, who would deny their child a marshmallow peep? We then graded our inventory for our partner feeding programs to show what was healthiest on a sliding scale. We hired a nutritionist-chef, taught about nutrition in the schools and feeding programs, recruited executives from the healthcare sector, launched a "food prescription" program, and set annual goals to distribute higher amounts of fresh produce. The food bank now has Mercy Kitchen, a separate facility that creates and delivers thousands of meals daily for Head Start programs, independent summer feeding sites for kids, and after-school programs.

In addition, the food bank launched a medically tailored meal program funded by the DaVita Giving Foundation. It provides free, fresh, prepared-from-scratch frozen meals tailored to low-income individuals suffering from advanced diabetes, kidney disease, congestive heart failure, COPD, hypertension, and HIV. These meals are delivered to homes, and services include nutrition counseling and education. A study of the effectiveness of medically designed meals showed they resulted in a 16 percent reduction in healthcare costs and 49 percent fewer hospitalizations.[17]

Along with many other food banks, we discovered that delivering more nutritious food enhanced our value to our community. After the food bank started distributing healthier foods, I received a call from the director of an after-school program in an elementary school where we provided food. Food was the magnet to get the kids' attention and draw them into the program. The director said, "The kids love the food." I replied I was delighted to hear that. She said, "The most amazing thing about the food was that most kids had not seen these foods before." The foods were baby carrots and cherry tomatoes with a few grapes. I recall another mission moment tied to one of our partner food pantries. The pantry volunteer came to our staff one day during a pickup and asked what was wrong with the rice; it was brown. We explained that brown rice is healthier than white, processed rice, that the brown color comes from the nutrients in the husk, and that it has to be prepared differently. So, there was a steep learning curve as we delved into proper nutrition.

In 2014, as part of that learning curve, the Second Harvest Food Bank of Central Florida convened a summit in Orlando on a food bank's role in community health. We invited interested colleagues from across the country, representing two major hospital systems, federally qualified health clinics, county health departments, higher education, health insurance companies, and select financial donors. A board member

---

17   Sarah Downer et al., "Food is Medicine: Actions to Integrate Food and Nutrition into Healthcare," *BMJ* (June 2020), https://www.ncbi.nlm.nih.gov/pmc/articles/PMC7322667; Seth A. Berkowitz et al., "Association Between Receipt of a Medically Tailored Meal Program and Health Care Use," *JAMA Internal Medicine* (April 2019), https://jamanetwork.com/journals/jamainternalmedicine/fullarticle/2730768.

from Advent Health emceed the summit and engaged the attendees in a robust conversation around "What is the food bank's role in community health?" Much ground was covered during that summit, and no immediate answers came forward; however, it prompted many important questions and potential ideas. Afterward, attendees started applying parts of the discussion to their local markets. In Orlando, we created a Health and Hunger task force in 2015, led by the food bank, that continues its work today. When the task force was established, it was comprised of twenty to thirty members representing the health community and higher education sectors and was led by Karen Broussard, VP of Community Impact. It took about a year and a half to gain a basic knowledge of what everybody brought to the table and find some commonality in our goals, service, clients or patients, and aspirations. Karen said understanding each other took a long time because of the differences in industries, business models, language, regulations, values, and priorities. She led the group in questioning the status quo of our respective work in particular areas and sought new solutions to some of the vexing problems surrounding people facing hunger.

One of the original task force members was Stephanie Garris, the executive director of Grace Medical Home, a nonprofit healthcare provider for uninsured individuals. It serves thousands yearly through the work of four hundred clinical volunteers, many of them retired physicians who want to give back, as well as medical students, social work interns, and a small but mighty staff. Stephanie says there is nothing more important than access to food. She emphasized, "We can't do health without food!" With a generous grant from the Orlando Magic Youth Foundation, the food bank and Grace Medical launched a program focused on 150 adults and 300 low-income, uninsured children. It centered around healthy food supply and nutritional counseling. The results were positive, with evidence of weight loss, fewer vitamin deficiencies, and overall improved health outcomes. Stephanie summed up the efforts of Grace Medical as follows: "We work with a challenging population, but we know how to do it. We need partners and resources, but we can't write off a small percentage of the population because it's too hard or they may be medically illiterate. People with low incomes are among us; let's take care of them. It's a sacred privilege."

Another member of the Health and Hunger task force, J. B. Boonstra, the director of Community Benefit for Advent Health, also saw the benefit of the collaboration. "The Health and Hunger task force was instrumental in changing the conversation among partners; it connected all of us. It provided immense value in exploring potential collaborative solutions to help people." Collaboration is one of many ways to impact someone's health before they get to the hospital. One example was integrating the Hunger Vital Sign screening tool to assess food insecurity in clinical settings. The tool has been endorsed by the American Hospital Association, the Academy of Pediatrics, and Feeding America, among other organizations. It is widely used in hospitals and clinics as part of a social needs assessment. The tool asks respondents to rank the following two questions as often true, sometimes true, or never true.

1. Within the past 12 months, we worried whether our food would run out before we had money to buy more.
2. Within the past 12 months, the food we bought didn't last, and we didn't have money to get more.

Some task force members have adopted this tool or are working on including it as a standard procedure. One question the task force members asked before adopting it was, "What if the patient does screen positive for food insecurity, then what? Where do we refer them?" That's where the partnership with food banks comes in—they can provide nutritious food and meals. This collaboration has started a popular "food prescription" model across the country. I consider this a big win.

Another outcome of the Health and Hunger task force was the creation of the Healthy Pantry Network pilot in 2017. Some of the five hundred food partners working with the Orlando food bank agreed to the following:

- Be open on days and hours of the week that are user-friendly.
- Distribute only fresh foods.
- Be a site for nutritional education.
- Accept referrals from clinics and hospital partners.
- Enroll recipients in a health assessment program that screens them for weight, BMI, diabetes, etc.

For partners enrolled in the program, hospitals and clinics provided trained staff for screenings, tests, and education and tracked the results of the screenings. Simultaneously, the food bank provided healthy fresh foods and communicated between the food provider and healthcare partners.

This program continues to offer effective interventions that prevent hospital admissions in many cases. The Healthy Pantry Network has demonstrated positive results in weight loss, BMI improvement, identification of diabetes or assessment of risk, and increased nutritional literacy. Iterations of this concept are being developed and operated nationwide in the food bank network.

Food banks continue to work with partners to improve nutrition while alleviating hunger. Meanwhile, the federal government recognizes the importance of healthy foods for low-income people. The White House held a conference in September 2022 to develop a national strategy on hunger, nutrition, and health. The executive summary reads, "The U.S. has yet to end hunger and is facing an urgent, nutrition-related health crisis—the rising prevalence of diet-related diseases such as type 2 diabetes, obesity, hypertension, and certain cancers. The consequences of food insecurity and diet-related diseases are significant, far-reaching, and disproportionately impact historically underserved communities. Yet, food insecurity and diet-related diseases are largely preventable if we prioritize the health of the nation."[18] I hope this innovative

---

18   "Executive Summary: Biden-Harris Administration National Strategy on Hunger, Nutrition, and Health," The White House, September 2022, https://www.whitehouse.gov/briefing-room/statements-releases/2022/09/27/executive-summary-biden-harris-administration-national-strategy-on-hunger-nutrition-and-health/.

approach to health and hunger spreads nationwide and realizes its full potential. Let's take advantage of the bounty of surplus crops available, eliminate wasted food, strengthen public–private partnerships, lower healthcare costs, and provide healthier lives for people facing hunger.

Job Training and Placement

Along with finding innovative ways to fight hunger, food banks nationwide have sought creative ways to help people escape poverty altogether. The food bank I led offers culinary training programs for low-income people, and this is true of many other food banks as well. The courses may differ a bit. However, they share similar elements. People who have shown a genuine interest can be expected to spend sixteen weeks in a commercial kitchen with professional chefs as instructors. They learn a variety of culinary skills, as well as participating in a parallel track for life skills such as household finances, interview skills, and resume writing. They receive support from a staffer who can help them with any outside issues they may be dealing with, such as finding an apartment, childcare, or a bus ticket, to mention a few. The course is offered at no cost to the student. The staff finds its graduates jobs that pay above minimum wage upon program completion, often with benefits. Statistics show that retention of these jobs is much higher than industry standards. Employers seek these graduates due to their training and healthy attitudes. Some graduates have started their own businesses. It's a win-win for graduates, families, employers, the community, and financial donors who want to promote economic stability.

Chef Kenny, a graduate of the Orlando program, shared his story of growing up in a low-income neighborhood and family, never going on a vacation, watching his cousin get killed, and having a long list of other experiences that were not happening in other areas of town. He developed a love of cooking at a time when he was struggling with employers who would continually break promises of long-term jobs. He had a side hustle selling boiled peanuts and flavored rice. He was married with two children, and the bills kept coming even when there was little or no income to pay them. Kenny remembered the year that he was laid off two weeks before Christmas. He had to either make a car payment, pay the rent, or get a few Christmas gifts for the kids. These tough choices were ongoing. He said those evenings at the dining room table looking at a stack of bills would make his chest tighten with fear. He wanted solid employment to ease the pressure, which could give him some freedom to spend time with his kids.

Kenny was introduced to the culinary program at the food bank through a friend and decided to try it. He graduated at the top of his class and was placed in a job at the most upscale country club in the area, a utopia for multimillionaires. He felt like a fish out of water when he pulled through the guard shack and into a parking lot filled with luxury cars; he wasn't sure he was allowed to park in the same lot. Kenny excelled in his role in the kitchen and occasionally had the opportunity to meet some residents. Being a cordial guy, he made some friends. One day, a resident approached him and said a friend would have a large event on his estate in the Atlanta area and asked if he would manage the gig. Kenny was paid nicely for the week in Atlanta and was provided airfare, a rental car, and a credit card with a $100,000 limit. The event was a huge success. He has since been invited back multiple times to cater more events. Kenny eventually started his own catering business; it wasn't a smooth road, but it was loaded with learning. Kenny said his best decision ever was enrolling in that culinary course.

The stories of transformation are bountiful. I remember the time one of the graduates came by the food bank to visit and said she had received a nice raise and promotion, which the team applauded. She received her first-ever income tax refund and donated it to the food bank out of gratitude. The team shed tears. Time and time again, we see these folks as "givers," not the "taker" stereotype that is widely held.

I salute chefs, and their strong work ethic. They're on their feet all day in a hot kitchen around fire and sharp knives, while meeting the customer's high expectations.

Despite the power of innovation in fighting hunger and its related problems, innovation is not the sole component of success. It is not always the silver bullet. That may sound contradictory to what I've been sharing in this chapter; however, there are times when innovation is not right and can be destructive. Stanford Social Innovation states, "Innovation is a complex process and depends on the unique constellation of many organizational and external factors in a particular context."[19] In other words, it isn't straightforward. Bad leadership, dysfunctional teams, and overambitious production goals all fight against innovation. Also, there is value in an organization's established core, the routine practices perfected over time, which shouldn't be ignored. If you are a leader, don't burn out your staff; they perform the day-to-day duties, and you are asking them to go above and beyond the regular call. Allow your team to recover. Leaders often get excited about change and don't realize its effects on the team that executes and delivers it. A friend said, "Don't boil the ocean; allow your team to adjust."

Sometimes, as leaders, we have to say no to some ideas, no matter how golden we think they are. Years ago, I heard of a few food banks that started grocery stores for low-income people in food deserts. What a great value these stores provide to a community! However, they typically require continual financial subsidies from the philanthropic world. My vision was to design an economically self-sufficient store as a social entrepreneurial venture. At face value, there was a strong rationale for the idea of providing low-cost, high-quality, healthy foods in a location accessible to many. Jobs would be created, and profits could be shared with the store employees and the community.

Such an endeavor would, however, be enormously expensive, as it would require buying land, building infrastructure, obtaining equipment, hiring employees, doing payroll, getting insurance, paying for goods, and dealing with slim profit margins and theft. I discovered that each food bank that attempted to start a grocery store of any scale lost money. The endeavor also demanded incredible time and energy, with little or no return. But I was stubborn, thinking I could

---

19   Christian Seelos and Johanna Mair, "Innovation is not the Holy Grail," *Stanford Social Innovation Review* (Fall 2012), https://ssir.org/articles/entry/innovation_is_not_the_holy_grail.

probably pull it off. I had a donor willing to provide the land, building, and capital infrastructure; with that under my belt, it would be much more manageable. I thought it was a no-brainer. Fortunately, I had the good sense to have a major consulting company conduct a feasibility study. Their report dashed my hopes and dreams; I said no to the idea. I almost became the dog in the analogy that caught the bus, now what!? Most likely, this was one of the wisest financial decisions I made over the years.

Chapter 10
Policy and Advocacy

*Every important change in our society, for good, at
least, has taken place because of popular pressure—
pressure from below, the great mass of people.*
- Edward Abbey

Food banks benefit from government funding that translates into a huge community impact. Pam Irvine, the president
and CEO at the Southwestern Virginia Second Harvest Food
Bank, said, "We must treat the government as our major
donor." A case in point was the $12 billion the American
Rescue Plan Act provided for nutrition assistance during
the COVID-19 pandemic. Tens of millions of dollars were
channeled through food banks. When a food bank looks at
its sources of revenues, the government stands out in terms
of providing funding and supplying critical food. Food banks
focus their advocacy efforts on the vital nutrition programs
that are in place in the United States.

Policy relating to nutrition programs has shifted over the
years, and food banks have weathered the storm, learning
advocacy along the way. Poverty was first measured in the
1960s; hunger was measured later. Poverty increased sharply
in the 1980s after Reagan-era cuts to several programs. In 1996,
during Bill Clinton's presidency, welfare reform was enacted,
with significant cuts falling on SNAP. That loss motivated many
anti-hunger and anti-poverty groups to rally and collaborate.
Advocacy among food banks started during this period. The
Bush administration replaced food stamps with Electronic

Benefits Transfer (EBT) cards and helped expand SNAP, just in time for the Great Recession of 2008, which saw an increased use of the program. Twelve years later, in response to the COVID-19 pandemic, the federal government increased aid to people in poverty, particularly with the Child Tax Credit, which decreased SNAP usage. The Child Tax Credit reduced childhood poverty by more than 50 percent.[1] However, Congress did not renew it in 2022, and consequently, poverty and food insecurity have returned to high levels.

Nevertheless, many federal politicians, lobbyists, and advocates have worked hard to address poverty in the United States, some starting as early as the 1960s, including high-profile people such as Bob Dole and George McGovern. I had the opportunity to interview Congressional Representative Jim McGovern (D-MA) for this book. As a senior member of the House Committee on Agriculture's Subcommittee on Nutrition and Oversight, Jim has tirelessly advocated for ending hunger in America and around the world. He fought for and secured the second-ever White House Conference on Hunger, Nutrition, and Health in 2022.[2] This conference put people with lived experience front and center to share their perspectives and input. He believes there is a big disconnect between his colleagues and the lives of people facing hunger. Most have never talked to a hungry person, and some members of the subcommittee don't even know the subject matter. Jim states, "Hunger is a political condition. Thirty-five members of Congress and 100 senators will never have to worry about going hungry."

Even though the amount of relief has declined, the federal government still provides substantial help to hungry Americans. This aid is funded through the Farm Bill, which is a significant focus for anti-hunger advocates. The bill

1   Annie E. Casey Foundation, "New Data Show that the Child Tax Credit Fueled a Substantial Reduction in Child Poverty," September 19, 2022, https://olis.oregonlegislature.gov/liz/2023R1/Downloads/PublicTestimonyDocument/61188.

2   "In Fight Against Hunger, McGovern Secures Historic White House Win," press release, Jim McGovern, May 4, 2022, https://mcgovern.house.gov/news/documentsingle.aspx?DocumentID=398857.

provides funding for forestry, research, conservation programs, crop insurance for farmers, and other categories. USDA's presidential budget request total for 2024 is $209.7 billion, with 7 percent going to commodities and 76 percent going to the nutrition programs that are administered by the USDA.[3] The USDA lists all of these on its website; here are the major ones:[4]

- SNAP (formerly called Food Stamps) provides families in need with help toward their food budget, enabling families to buy healthier options and start the transition out of poverty.
- TEFAP provides free emergency food assistance to low-income Americans and funds to the states to administer the program.[5] Over 50 percent of these foods are distributed through the Feeding America network and are a large, vital part of the network's distribution.
- The National School Lunch Program, started in 1946, and the School Breakfast Program, started twenty years later, provide free or low-cost nutritional meals to children in need.[6]
- WIC helps low-income pregnant women and children up to five who are at nutritional risk. It does this by providing funds to states that pay for health care, education, and nutritious food.[7]

---

3 USDA, *Budget Summary FY2024,* Usda.gov/sites/default/files/documents/2024-usda-budget-summary.pdf

4 "Food Assistance Programs," Nutrition.gov, USDA, 2023 https://www.nutrition.gov/topics/food-security-and-access/food-assistance-programs.

5 "The Emergency Food Assistance Program," Programs, USDA, 2023, https://www.fns.usda.gov/tefap/emergency-food-assistance-program.

6 "National School Lunch Program (NSLP) Fact Sheet," updated November 2017, https://www.fns.usda.gov/nslp/nslp-fact-sheethttps://www.fns.usda.gov/nslp/nslp-fact-sheet; "SBP Fact Sheet," updated November 2017, https://www.fns.usda.gov/sbp/sbp-fact-sheet.

7 "Special Supplemental Nutrition Program for Women, Infants, and Children," WIC, USDA, updated February 5, 2024, https://www.fns.usda.gov/wic.

- The Child and Adult Care Food Program (CACFP) reimburses daycare centers for children and adults, after-school care, and emergency shelters for the cost of providing healthy meals and snacks to eligible children and adults.[8]

Much of the funding in the Farm Bill goes to SNAP, which is the most effective tool for fighting hunger. However, misunderstandings and myths exist around the program. The majority of SNAP recipients do not work because they cannot. In 2019, 43 percent of recipients were children, 16 percent were elderly, and 10 percent were disabled.[9] Most of the remaining SNAP recipients work.[10] SNAP is not a charity; it is available to all eligible citizens.[11] Nor does SNAP completely cover needy people's grocery bills. The average benefit works out at about $4.20 per person per day.[12] Additionally, SNAP's benefits extend well beyond the program's recipients. The money from SNAP is spent in local stores, benefiting the economy more broadly. A study in 2019 found that every dollar provided in SNAP funds generated $1.50 of GDP for the nation.[13] Finally, helping people eat enough healthy foods reduces healthcare costs. A study published in 2017 found that "SNAP participation was associated with approximately $1,400 per year

8   "Child and Adult Day Care Food Program," USDA, updated February 05, 2024, https://www.fns.usda.gov/cacfp.

9   "Characteristics of SNAP Households: FY 2019," USDA, March 29, 2021, https://www.fns.usda.gov/snap/characteristics-snap-households-fy-2019.

10   "The Relationship Between SNAP and Work Among Low-Income Households," Center on Budget and Policy Priorities, January 30, 2013, https://www.cbpp.org/research/the-relationship-between-snap-and-work-among-low-income-households.

11   "5 Myths About SNAP Food Assistance Benefits," Vantage, December 27, 2019, https://vantageaging.org/blog/myths-about-snap/.

12   Ellen Vollinger, "Thrifty Food Plan Revision Is Welcome News," Food Research and Action Center, October 1, 2021, https://frac.org/blog/thrifty-food-plan-revision-is-welcome-news.

13   Patrick Canning and Brian Stacy, "The Supplemental Nutrition Assistance Program (SNAP) and the Economy: New Estimates of the SNAP Multiplier," Economic Research Service, USDA, July 2019, iii, https://www.ers.usda.gov/webdocs/publications/93529/err-265.pdf.

per person lower subsequent health care expenditures on low-income adults."[14] Additionally, increases in SNAP benefits decreased hospitalizations among Medicaid recipients, which promptly rose again once the benefit was cut in 2013.[15]

Beyond these specific nutritional programs, two other federal policies have significantly helped hunger relief efforts. They were created and passed in the earlier days of food banking; one protects companies from legal liability connected to their donations, while the other relates to tax incentives to corporations to donate food. The first policy, the Bill Emerson Good Samaritan Food Donation Act of 1996, provides limited liability protection for individuals and organizations, from farmers to restaurants to manufacturers, who make good-faith donations to nonprofits that feed people facing hunger.[16] Every state has similar legislation that limits donors' liability.[17] The second, passed in 2011, is the Internal Revenue Code 170(e) (3). It was designed to encourage businesses to donate food to food banks and other nonprofit organizations. It allows businesses to deduct the cost of the food from their taxes, as well as half the value of a reasonable profit if the food has been sold. In December 2015, the Protecting Americans from Tax Hikes Act extended these tax deductions to more companies and made them permanent.[18]

14   Seth A. Berkowitz et al., "Supplemental Nutrition Assistance Program (SNAP) Participation and Health Care Expenditures among Low-Income Adults," *JAMA Internal Medicine* (November 2017), https://www.ncbi.nlm.nih.gov/pmc/articles/PMC5710268/.

15   Rajan A. Sonik et al., "Inpatient Medicaid Usage and Expenditure Patterns after Changes in Supplemental Nutrition Assistance Program Benefit Levels," *Preventing Chronic Disease: Public Health Research, Practice, and Policy* (October 2018), https://www.cdc.gov/pcd/issues/2018/18_0185.htm.

16   "Frequently Asked Questions about the Bill Emerson Good Samaritan Food Donation Act," USDA, 2022, https://www.usda.gov/sites/default/files/documents/usda-good-samaritan-faqs.pdf.

17   Jean Buzby, "Federal Incentives for Businesses to Donate Food," USDA, July 8, 2020, https://www.usda.gov/media/blog/2020/07/08/federal-incentives-businesses-donate-food.

18   "Donations," USDA, 2022 https://www.usda.gov/foodlossandwaste/donating#:~:text=Enhanced%20tax%20deductions&text=Qualified%20business%20taxpayers%20can%20deduct,value%20of%20the%20donated%20food.

Despite this federal aid, millions of people in the United States are still hungry. This challenge gives focus to the efforts of hunger relief champions and food bank advocates. Why don't we fund anti-hunger programs and eradicate hunger once and for all? Why do our policies fail to support this goal? The United States has by far the largest economy in the world, with a GDP of $26 trillion. John Sayles, CEO of the Vermont Food Bank, recalls a comment he heard at a national food bank conference from an Illinois commodity farmer. The farmer's name is Howard Buffet, and he is one of the sons of Warren Buffet, the multibillionaire investor. Howard said, "There is plenty of food and money; all we lack is the political will to end it." Similarly, Vince Hall, the chief government relations officer for Feeding America, told me in our interview, "We set up fire stations all over the United States at taxpayer expense to put out fires. Well, hunger is a raging fire, a raging fire across our country. Can you believe we have public libraries giving people internet access to books and access to learning, but we don't have a policy that makes sure they have enough food when they go into the library?" Our policies are at the heart of the problem. Hunger isn't just a failure of public policy. It's a failure of public virtue. Advocacy must shine a light on this issue.

A significant milestone for Feeding America was reached in the mid-1990s when the network got involved with advocacy to influence and defend policy. I'll never forget one national food bank conference where Sr. Christine Vladimiroff, then-CEO of Second Harvest Foodbank Network, announced that the national network would get involved with advocacy as part of the mission. Most food bankers, including me, pushed back hard on the idea. We said our business was moving food in and out of our facilities and feeding people, not advocacy. We were so involved in the operations side of food banking that we could not see the opportunity to use our connections and knowledge, help educate others, and speak on behalf of those we served. Unsurprisingly, Sr. Christine, a Catholic nun, was the one who made that historic moment; it was all about social justice. Fast forward to today, and you'll find that just about every one of the two hundred members of the network is actively involved in advocacy to some degree. Before that, we had not realized the power we had to become a voice for those facing hunger. I became a strong advocate after seeing the light.

Many other groups have been instrumental in developing effective advocacy for policy change, such as the Food Research and Action Center (FRAC), Bread for the World, the Center on Budget and Policy Priorities, No Kid Hungry, the Alliance to End Hunger, Catholic Charities, and MAZON: A Jewish Response to Hunger, to name a few, as well as myriad regional, statewide, and local entities. When we combine our voices, much more power is brought to bear.

Individuals are also significant. Doug O'Brien, now vice president of Programs at the Global Food Banking Network, was Feeding America's first advocacy and government relations hire. He was instrumental in establishing the national network's priorities, engaging the membership, educating legislators at the federal level, and establishing a critical relationship with the USDA. Doug personifies the qualities of an advocate: courage, independence, and passion for the things that matter. Doug's effective work, and that of others, is recognized by the federal government. Eric Bost told me, "Feeding America is a strong advocate for people facing hunger and [for] nutritional programs in every part of the country." This sentiment was echoed by Stacy Dean, the current deputy undersecretary for the USDA's Food, Nutrition, and Consumer Services, who said, "Feeding America is an excellent partner with a unified voice, solid analytical prowess, and a partner in disaster relief and crisis." She values the spotlight on issues shared with her and how Feeding America shares best practices.

One of the powers of Feeding America food banks is that they touch people in every congressional district in America. Rob Fersh is a former congressional staff member and president of FRAC, a leading national organization working to improve public policies to eradicate hunger and undernutrition in the United States. He told me, "Since food bankers are so embedded in their communities and have such a close connection with the people, they serve, they have a special credibility with policymakers beyond what many advocates have." Ellen Teller, the FRAC's chief government affairs officer, states that she is proud of the collaboration with Feeding America, and a great example of this is their cohosting of the annual National Anti-Hunger Policy Conference. Over a thousand advocates gather annually in Washington, DC, to deepen their knowledge of current policies, best practices, and communications strategies. During the conference, time is set aside to visit with elected officials.

I want to share a few stories about practicing advocacy. Here's one on what not to do. I'll never forget my first rookie trip to my state capitol years ago as part of a larger group advocating for the same issue. We sat down with our first elected official, and our group leader launched a full attack. I just wanted to slide under the table and disappear. My experience was a mixed bag of emotions. I found that advocacy was very hard work and, at times, very rewarding. I experienced frustration, anger, cynicism, and disbelief when studying policy recommendations that seemed oblivious to the facts. Walking and working the halls of Congress is a roller coaster for me. Many fine people are serving us; however, they're caught up in a dysfunctional system. Knowing those fine folks gave me hope because I couldn't imagine the outcome without them. It seems like so many decisions are made based on extreme ideology, warped theology, personal likes and dislikes, and ego. Thank goodness for some sage words from Leonard Cohen, the Canadian singer-songwriter, poet, and novelist. I had a plaque that was on my office wall that reminded me daily of his advice; it is from a song called "Anthem."

> Ring the bells that still can ring
> Forget your perfect offering
> There is a crack, a crack in everything
> That's how the light gets in.

Tell the story you can tell and keep telling it. Look for that crack.

I will share a story of a "crack in the bell" that I experienced. It was at breakfast in 1994 with one of my board members,

Representative William Lehman. He served in the US House in the 1980s and 1990s and represented Miami's 13th and 17th districts at different times. We had breakfast at a local Miami Beach diner that had been around for decades. The congressional representative, originally from Alabama, had his stash of grits in their fridge. In our conversation, I was sharing my concerns about funding for hunger relief at the national level. I hoped he could advise me on how to reach out to Washington, DC. I had never been to the capital before, and the congressman asked me if I wanted to go. I readily accepted the invitation, and he asked whom I wanted to talk to.

I immediately said the secretary of agriculture, who is responsible for federal funding for nutrition programs. In addition, I want to meet with a few politicians who did not want to support our cause. Bill Lehman made it happen. As we walked the halls of the Capitol, it was clear that Bill was a first-class guy; everyone had the most profound respect for him from either side of the aisle. Conversations with both Republicans and Democrats were civil. That trip to Washington included a conversation with the secretary of agriculture at the time, Dan Glickman (and his beagle that roamed freely throughout the offices). Eventually, it led to an opportunity to testify to Congress twice (for me, not the beagle) on behalf of perishable food recovery programs and the potential to gain millions more meals. That breakfast had cracked open the door!

Similarly, while visiting a legislator who represented my state in Washington, DC, I showed up for my appointment and was told I had to meet with a legislative aide, not the senator; this is always disappointing but not uncommon. I eventually learned that these aides are essential to know. Anyway, I told the aide about the hunger issue and what we were doing about it. It was a controversial policy issue, and I wanted to see how the senator felt about it. I received a polite non-answer. The aide kept looking at her watch and finally said that other matters required her to end our discussion; she seemed disinterested.

As I was about to leave, I handed her a food bank brochure, and she quickly flipped through the pages, trying to appear interested. She stopped, staring at a page, then looked up at me and said, with a tear in her eye, "Do you provide food to this charity?" She had pointed to one of the random examples we had included about our five-hundred-partner network, a program for troubled juveniles. She said that her brother had been in that program, which had changed his life. She had never tied the fact that food is essential to the mission of that program. That "crack" totally changed her demeanor and attitude toward our message. This was the first step in establishing a connection with the senator.

Another example of finding that crack happened in Orlando when I was meeting with elected officials and getting to know what was most important to them. It's also a story of how viewpoints that seem so far apart can be so close in other ways. I invited a conservative congressional representative to the food bank for a tour and a conversation about SNAP funding concerns. When I brought up the topic, he quickly dismissed my point and said, "I have heard enough of it already from others." The tone of voice and body language were formidable. I paused and practiced the ten-second rule without saying anything and gathered myself.

I led him to our community kitchen, where the culinary job training and job placement program is located. He witnessed low-income and homeless students learning a new skill and preparing for a career in food service. Upon completing the sixteen-week program, the food bank places graduates into jobs. I explained folks were becoming taxpaying citizens who could now put food on their tables. The congressional representative's entire countenance changed as he spoke to the students about their experience, what they were learning, and their hopes and dreams. He was all smiles and asked to be the commencement speaker at the next graduation ceremony.

What's not to like? Everyone can agree on economic development, and to people learning new skills, getting jobs, being great examples to their families, and having dignity and hope for the future. While wrapping up our discussion for the day, I couldn't resist bringing the conversation full circle. I took a chance and let him know that all of those students needed SNAP for a short period until they could get back on their feet. He was undoubtedly a new friend of the food bank.

However, the way to connect with elected officials in Washington, DC, is not always obvious and can be challenging. There is often little time to speak to them or their legislative aide, and you might wait while they're called in to vote on an issue. When you walk into an office, you're in a small entrance area with people standing around for their appointments. The phone is ringing off the hook from constituents across the country, and there are tall stacks upon stacks of printed material and bills to review. Your issue may be number ninety-nine on their list. You may sometimes meet with a legislative aide, not the representative or congressperson. However, you must hang in there and tell your story despite aspects that may seem challenging. I finally learned during my years in food banking that advocacy is a marathon, not a sprint. Significant changes often take time. The typical nature of food banking is to search for an instant fix to everything and to work with an incredible sense of urgency. Food banks have an untethered organizational structure that is nimble, flexible, and unbureaucratic. So, it's a culture clash when walking the halls of power in Washington, DC.

Ellen Teller told me one story about how advocacy for after-school feeding programs arose in the 1990s. She recalled a high school basketball coach who visited her in Washington, DC. Together, they lobbied with the Police Athletic League. The coach said, "You know my boys, they want to play hoops and shoot hoops after school, and I could run an after-school basketball program. The program could become viable if the federal government reimburses us with these snacks and meals for low-income children and families." The coach said he would offer the meals, and then before the guys could play ball, they would have to do a mandatory forty-five minutes with their mentor. He would get the community college students to come in, set up a program with the kids, assign a homework assistant, or whatever was needed. The mentors would then sign an index card saying the kids had been there for forty-five minutes and covered their assignments. Then they could compete, play ball, get their snack and meal, and go home. They got some nourishment. They could be less antagonistic with their parents because they weren't hungry, had their homework done, and had fun. The other benefit of this program, which continues today, is that it keeps youth off the streets during the midafternoon hours when bored minds are dangerous. Violent crimes committed by youth most often happen immediately after school.[19] Not only do the students benefit, but also the family and the community.

Feeding America established the Policy, Engagement, and Advocacy Committee (PEAC) in 2011, which is the national network's government relations committee. It aims to create a more systematic, consistent, and strategic way to engage network members in public policy and advocacy efforts. It comprises a select small group of network CEOs and Feeding America staff. I enjoyed serving on the committee and found the two-way communication very effective. The national team shared their knowledge and perspective, while food banks offered their local perspective. Through collaboration, we developed priorities, strategies, and messaging around influencing federally funded nutrition programs. This approach is a vehicle for bringing two-hundred-member voices to the national table. Individual and collective advocacy took place in a highly coordinated manner. Carrie Calvert, VP of Government Relations, Agriculture, and Nutrition for Feeding America, believes that despite the massive success of the network's efforts, food banks still do not realize the full power of working together. However, there is momentum in the right direction for it to become even more of a force for people facing hunger. Due to the collaborative structure of the PEAC, hundreds of millions of dollars have been secured and protected for nutrition programs.

One of the most memorable PEAC meetings invited Democratic and Republican communications consultants. Individually, they presented the "temperature" on the Hill regarding our issues. Their input was invaluable to hear when considering our messaging. It may have been the same meeting where Feeding America invited a linguist. This presentation focused on the type of language to use when speaking to a

---

19   "Statistical Briefing Book," Office of Juvenile Justice and Delinquency Prevention, 2022 https://www.ojjdp.gov/ojstatbb/offenders/qa03301.asp.

Republican or Democrat. It was fascinating to hear the dos and don'ts. Words carry much weight, and one word can either relate to another person or alienate them. The presentation covered individual phrases and words and how to structure a conversation with an elected official.

Here's a small sample of what we learned. When meeting with someone not in favor of your position, do not start by presenting them with a problem; they have enough of them on their desk. Do not become combative, or you'll build a wall between them and you. Start with establishing some common ground. Steer toward solutions, working together. This kind of shared knowledge is essential, especially when you want to maximize your limited time with an official. Sabeen Perwaiz, president and CEO of the Florida Nonprofit Alliance touted the positive approach: "They are people first; thank them for their service. Know how your impact statement will affect their bottom line and localize the information to their area." Jim McGovern echoed this advice, "You must understand where your member of Congress stands and hold them accountable for what they're doing. You have to fight for what you believe in."

Nonprofits must advocate because the political system has been significantly influenced by big money and special interests with many voices at the table. In 2021, corporate America spent $4.09 billion on lobbying.[20] Every conceivable entity is involved with advocacy, ranging from gun control and pharmaceuticals to casinos, tobacco, health care, and everything in between. Nonprofits don't make financial contributions to political parties, candidates, or causes, so it's harder to be heard through the myriad of other efforts and influence elected officials. The advocacy of nonprofits is generally centered on social justice. We want to change the "what is" into "what should be," namely a more decent and just society. Establishing relationships with elected

---

20   "Total Lobbying Spending in the United States from 198 to 2022," Statistica, 2023 https://www.statista.com/statistics/257337/ total-lobbying-spending-in-the-us/.

officials is essential at several levels, from city or county officials to state and federal positions. Robin Safley, the executive director of Feeding Florida, a statewide association of Feeding America food banks, astutely notes the role of food banks, or most nonprofits in general: "Politicians listen to two types of people. The person who sits beside them in church or gives them money. You need to be the person who sits beside them in church."

Here's an example of sitting next to them in "church." A few of us from the food bank were in Washington, DC, making the rounds with elected officials from our district in Florida. Although we were only metaphorically "sitting in the pew" with them, we localized the information and made it as personal as possible. We presented food insecurity and poverty information for their specific district, what we were doing about it, and what needed to be done. It was a person-alized picture of their backyard. After the presentation, the representative turned to her legislative aide and said, "They just saved you hours of work. We now have a much better idea of what's going on." We became their trusted partner.

When I led the food bank in Orlando, the board and I made a more profound commitment to advocacy. After establishing our goals, it was clear that we needed to create a full-time position for advocacy. I recall interviewing several well-qualified candidates for the job. During an interview, one candidate, Kelly Quintero, shared that when she was in college during the Great Recession and living at home, her parents went through a tough time and found their finances limited due to unemployment. Their family depended on SNAP to bring food to their table. Like many other SNAP recipients, their need was temporary to get them through a challenging period. Kelly owned the issue in her heart; I hired her immediately. As the director of Advocacy and Government Relations, Kelly was a person of lived experi-ence and had a unique energy and passion for a cause. She is now a policy director for Feeding America's Washington, DC, office. Kelly's story illustrates one of the most potent ways of educating an elected official—by having someone impacted by an issue speak out.

Claire Babineaux-Fontenot, CEO of Feeding America, emphasizes this approach through an initiative called Elevating Voices to End Hunger Together. The initiative overview states, "To end hunger, we must listen to the people experiencing it."[21] Feeding America is gathering input from people with lived experience in various ways, including insights from nearly thirty-six thousand people facing challenges getting the food they need. The anti-hunger policy recommendations—informed by these people—prioritize dignity, increase access, expand opportunity, and improve health.[22]

Here's another great example of how a person's voice can impact decision-makers, even at the highest level. Storytelling has much power. Claire shared a decisive moment from the most recent White House Conference on Hunger, Nutrition, and Health, held in September 2022. The conference's goal was to "accelerate progress and drive significant change to end hunger, improve nutrition and physical activity, reduce diet-related disease, and close the disparities around them."[23] In an activity before the conference, Claire had invited some women of lived experience to share their stories with the president. Anita, one of the women, shared her aspirations for the conference and spoke about what mattered most. During the conference, Claire and Anita sat in the last couple of rows in the auditorium. The president gave his welcome address, and after he spoke, Anita smiled and said, "Those were my exact words from the pre-conference meeting with him, and I'm hearing the President of the United States say them right now!" As Claire told me, "This lady who has been food insecure generationally, who suffered through so much

21   "Elevating Voices to End Hunger Together," Feeding America, 2022 https://www.feedingamerica.org/research/community-solutions-hunger.

22   "Executive Summary: There is a Crisis in America," "Elevating Voices to End Hunger Together," Feeding America, https://www.feedingamerica.org/research/community-solutions-hunger. 2022

23   "White House Announces Conference on Hunger, Nutrition and Health: Government Relations & Washington Update," Agricultural and Applied Economics Association, May 2022, https://www.aaea.org/publications/the-exchange/newsletter-archives/volume-44---2022/may-2022-issue-10/government-relations--washington-update.

drama and trauma throughout her life, who saw her mother struggle, who saw her grandmother struggle, this woman, Anita, heard the President of the United States of America say something that she told him. It was beautiful. I'll never forget that day or that experience."

Advocacy and food-focused government policy are necessary. Charities alone cannot end hunger. Feeding America member food banks distributed six billion pounds of food in 2021; despite that herculean effort, an enormous gap still existed to fill the need. Federal nutrition programs provide roughly ten times as much food assistance as private churches and charities. As Stacy Dean of the USDA points out, "There must be a broad approach to solving hunger: government, business, citizens, faith communities, and food banks. The government will always be at the core of the effort." She says that hunger is a downstream issue, and upstream investments must be made; a more systemic solution must occur to end hunger. The government must provide a baseline of support while food banks are needed to fill gaps that the government can't handle, especially when emergencies arise. Both private nonprofit and public programs are required to meet the need. Solving hunger is also about justice and making the United States a more just nation.

The best hope of ending hunger is to get together and demand it is done in our country. David Beckmann, former president and CEO of Bread for the World, said we have hunger because of "a lack of organized give a damn, particularly at the political level." John Sayles states that hunger is not a partisan issue but a lack of collective political will. The United States has the means to end hunger. We must find the will to do it—not reduce hunger but end it. Kate Maher, CEO of the Greater Chicago Food Depository, summed up my feelings and those of many others when she told me, "I don't want more food pantries; I want more justice!"

Chapter 11
Leadership, Culture, and Vision

> *The health and success of an organiza-*
> *tion or company rest on a three-legged*
> *stool: Leadership, vision, and culture.*
> **- Yours truly**

As we have seen, innovation, policy, and advocacy are important tools in the fight against hunger. However, none of these will help if the nonprofits on the front lines are poorly run. Organizations or companies succeed because of many factors. However, there are three especially powerful elements: leadership, culture, and vision. When these elements are aligned and working together, remarkable results happen.

Leadership
I've been a student of leadership throughout my three decades in food banking: reading, learning from others through observation, benefiting from mentors, and doing my best through daily practice. Through my reading, study, and the one hundred interviews I conducted across the country for this book, I heard about a variety of crucial leadership qualities. I will focus on a select few that I believe are especially important: self-awareness, attitude or mindset, possibility thinking, and approaches to failure.

The quality of being self-aware is featured in Jim Collins's monograph *Good to Great and the Social Sector.*[1] His book identifies

---

[1]   Jim Collins, *Good to Great: Why Some Companies Make the Leap and*

five levels of leadership based on research on the most successful companies over the decades; the highest level includes self-awareness. The levels range from capable to "Level 5: Executive," at which a leader "builds enduring greatness through a paradoxical blend of personal humility and professional will. This ties into the fact that a great leader must be self-aware to accomplish such a level."[2]

Mark Hertling also emphasized the importance of self-awareness. Mark mentioned there are three critical elements of leadership: First, who are you? Do you understand your attributes and competencies? In other words, do you have strong self-awareness? When it comes to your presence, are you who you are thought to be? Do you walk the talk? Second, what is your ability to influence people? How do you inspire others? Third, you must understand your working situation; it requires adaptation. In addition to those three key elements, he stated you need to be able to work with the higher-ups and the folks on the street level. Be a good communicator; one of the reasons some companies fail is poor or little communication. Intellect and IQ alone are not enough. Be aware of what you know—and don't know. Then realize that you don't know, and desire ongoing learning.

I want to share a personal story about self-awareness and its importance. I had the good fortune years ago of attending a weeklong course at the Center for Creative Leadership in Greensboro, North Carolina. Before the session, participants took a few lengthy surveys on our business experience and leadership style and had a 360-degree performance evaluation done. So, it was not only our supervisor who rated us, but also our peers and team members. This type of evaluation can be powerful in identifying how well you "know thyself" beyond your echo chamber. We all have our blind spots, strengths, and weaknesses and may not be aware of them. About forty professionals from various business sectors, higher education, and the nonprofit world were brought together; none knew each other. We were tasked

Others Don't (Harper Business, 2001).

2   Jim Collins, *Good to Great and the Social Sectors, A Monograph to Accompany Good to Great* (Harper Collins) October 6, 2001

with running a fictitious company over the following few days, and each person was assigned a role. Mass confusion ensued. Psychologists observed our activity individually and collectively during the next few days.

So much learning came out of this exercise that it profoundly changed my professional life. Session leaders addressed us on the final day and let us know how we performed as a team and individually. The real enlightening moment for me was when I received the report they provided, which was packed with an analysis of our individual behavior during the session, combined with the 360-degree survey results. The report was sorted into four buckets, as follows: Number one, things you're good at and you know it. It's always a good reminder to continue to do these things. The second bucket was things you're good at and didn't know; that's very affirming and encouraging. The third bucket was things you're bad at and know—a grim reminder to change things. The fourth bucket was a showstopper—something that you're not good at and didn't know, your blind spot!

This whole analysis and reflection focused on self-awareness. The session facilitators recommended sharing these discoveries with our teams back at the office. My blind spot was welcomed with much laughter. What was it? Lack of delegation. That discovery was a substantial advancement in my professional development and leadership. What I learned was to provide clear goals and expectations to staff, and then get out of their way and let them do their jobs.

Leadership is also a mindset or attitude. Leadership mentality is so critical. It can inspire, or it can discourage. As a leader, remember that people are always watching. You can influence others, whether you're a parent, teacher, supervisor, manager, or CEO. I benefited from some of the tips created by the Conscious Leadership Group. One is "Above the Line vs. Below the Line" thinking. It's shifting between "to me" and "by me" worldviews. In a document provided by the Conscious Leadership Group, there is a horizontal line drawn at the midpoint of the page. Above the line are beliefs and behaviors that involve presence, curiosity, growth, and learning. Below the lines are thoughts and behaviors that deal with drama, defensiveness, and scarcity.[3] I found it a simple but powerful tool to review occasionally as a personal check-in on my headspace and attitude. I used it often as a personal mantra and reminded myself to stay above the line.

Over the years, I have found that many organizations and people operate from a scarcity mindset. This mindset immediately and automatically creates an uphill battle, a negative outlook, and a lack of hope. It's almost a fearful approach. I'm an advocate of creating a framework of possibilities. This opens a positive, creative, exciting, and energizing headspace. This framework is at the heart of any healthy organization, company, or personal outlook. It's about imagination, not being locked into the status quo. Admittedly, there is great power in the status quo—think about examples in your life for a few minutes. I'm not advocating for total disruption or tyranny; it's essential to acknowledge that we need structure around us. However, there are plenty of opportunities to move beyond just what's in front of us. Thank goodness for Rosalind Brewer, the Wright brothers, Claudia Jones, Jane Cooke Wright, Benjamin Franklin, throughout history. Marianne Williamson captures this sentiment wonderfully: "The key to abundance is meeting limited circumstances with unlimited thoughts."[4]

The most valued book I've read that has been my go-to, professionally and personally, is *The Art of Possibility*. I have reread it numerous times and have gifted it over the years. The book is jam-packed with wisdom. Below, I include a portion of a chapter that focuses on a critical point of mindset, one I value deeply as a leader.

> The leader of the possibility invigorates the lines of affiliation and compassion from person to person in the face of the tyranny of fear. We can exercise this kind of leadership, whether we are CEO or employee, citizen or elected official, teacher or student, friend,

3   "Locating Yourself: Above or Below?" Conscious Leadership Group. Concencious.is/exercises-guides/locating-yourself-above-below.

4   "Marianne Williamson Quotes," Goodreads, from *A Return to Love: Reflections on the Principles of "A Course in Miracles,"* update or access date, URL.

or lover. We are living in the land of our dreams. This leader calls upon our passion rather than our fear. She is the relentless architect of the possibility that humans can be. Leaders who become their vision often seem uncommonly brave to the rest of us. Whether from the middle of the action or the sidelines, they are a conduit for carrying the vision forward.

The practice of framing possibility calls upon us to use our minds counterintuitively: to think about the contexts that govern us rather than the evidence we see before our eyes.[5]

I coach small nonprofits and often see a mindset of scarcity. To encourage them to embrace possibility, I ask them, "What would you do if someone walked into your office and asked what you would do with a million dollars?" Typically, in their response, you can see a change in their energy when they start dreaming about what could be done. Their enthusiasm immediately starts to shift into a higher gear . . . one of possibilities.

Another perspective on possibility thinking comes from a friend and colleague of mine, Mary McBride. She is chairperson of Design Management and Creative Enterprise Leadership at Pratt Institute, as well as a consultant to enterprise leaders worldwide. Her most recent book is *Leading as If Life Matters: An Invitation to Attend a Future of Our Own Making.* The concept of leading as if life matters is compelling to me and implies much responsibility; it conveys a moral imperative to carry on. Mary states, "Line up your thinking, knowing, doing, and being. Once you do proceed, you don't know until you go. Do whatever it is and discover who shows up and what resources do. The energy will support you."[6]

A real-life example of how food banks applied the quality of possibility thinking was with COVID-19 relief. The food

5   Rosamund Stone Zander and Benjamin Zander, *The Art of Possibility: Transforming Professional and Personal Life* (Penguin: 2002), 178.

6   Mary McBride, Maren Maier, and Xue Bai, *Leading as If Life Matters: An Invitation to Attend a Future of Our Own Making* (independently published, 2021).

bank I was leading, Second Harvest Food Bank of Central Florida, is based in Orlando. When Disney World shut down in mid-March 2020, it dawned on me that this virus was severe. Then all the other major attractions, like Universal Orlando and others, closed shop. The wave of job losses was like a tsunami. Restaurants, hotels, car rental companies, the convention center, and hundreds of related businesses were closed. That wave impacted hundreds of thousands of white- and blue-collar jobs. Multiply the Central Florida impact by millions to understand the national implications. If there was a time calling for solid leadership, this was it.

Food banks across the country braced themselves and responded in heroic fashion by being adaptable and flexible. Thousands of calls came from people asking for food; many had never been in that situation. Programs had to be shut down and new ones started amid all the craziness. Mark's statement about the importance of knowing one's strengths and attributes and being self-aware was essential for performing during this disaster. It was critical to recognize one's potential blind spot and acknowledge personal strengths and weaknesses. Relying on others who may know more than oneself and trusting they would carry out the vision were imperative.

So many resources were needed to respond; this is where that mindset of possibility came into play. Mary McBride's advice from our interview was right on point: do whatever it is and find out who shows up and what resources become available. The energy will support you.

The generosity of people across the country was outstanding. In my years of food banking, no other response has even come close to the outpouring from friends and strangers alike. Food donations and financial support showed up. Both local and federal governments provided financial resources and food at unprecedented levels. Corporate sponsorship, public and private foundations, and thousands of individual donors all came to the table, each playing a critical part in the solution. Food banks shifted their business model to mass distribution due to the staggering volume of people needing food. To maintain social distancing, cars and vans pulled into loading areas such as huge parking lots in stadiums and other public spaces, and volunteers loaded their trunks with food. Food banks saw their distribution volume increase 100 percent overnight. In Central Florida, we distributed enough food for 250,000 meals per day. So much of this impact was due to the attitude of possibility. The community showed up with generous support for food and funds.

Failure

An important quality of leadership is a nuanced attitude toward failure. In this section, I'll address failure: how it can have an upside, its conditions for acceptance in an organization, and some examples. I believe failure is often feared, hidden, or avoided, and I wish to shed some light on it. Ellen DeGeneres once said, "It's failure that gives you the proper perspective on success."

Failure occurs in many instances; it's not a stranger to anyone. Denise Parris, a wise and insightful friend, elite athlete, and entrepreneur, said, "To be alive means you're going to fail." So, let's not try to be perfect, which is unattainable anyway, and get along with life. It's a path to success. "If we were not provided with the knack of being wrong, we could never get anything useful done. We think our way along by choosing between right and wrong alternatives, and the wrong choice must be made as frequently as the right one. We get along in life this way. We are built for mistakes, coded for error. We learn, as we say, by 'trial and error.' Why do we always say that? Why not 'trial and rightness' or 'trial and triumph'?"[7]

Fear of failure is ingrained into us from a very early age; we define ourselves by outcomes. Remember when you came home from school and heard questions like "How did you do on your test? Where's your report card? Did you win the game? What was the score? Did you score?" Sadly, some people's childhood experiences are so traumatic that they carry them into adulthood. Did you ever try to give some constructive criticism to someone, and they freaked out over something seemingly small?

---

7   Lewis Thomas, *The Medusa and the Snail*, Published in Penguin Books, January 1, 1995, Penguin Random House

Failure is always associated with losing; we're taught that losing is bad. It's baked into education, sports, and careers. It's pervasive and made fun of. A friend of mine, Rob Brodnick, comes from higher education and is an author, facilitator, and business consultant. He says, "It's a winner/loser dichotomy. It comes with a cultural and social stigma that makes you feel less human." John Kennedy said, "Success has many fathers, but failure is an orphan."

Society views failure negatively and is set up only to reward success. Our society worships successful people: Super Bowl MVPs, winners at the Oscars and the Golden Globes, and many others.

The fear of failure can have a paralyzing effect; as a result, it inhibits any chance of progress. J. K. Rowling summed this up well: "It is impossible to live without failing at something unless you live so cautiously that you might as well not have lived at all—in which case, you fail by default."

"Failure is not always bad. Organizational life is sometimes bad, sometimes inevitable, and sometimes even good. Learning from failure is anything but straightforward." Failures differ widely; in many cases, they are considered positive because they provide new knowledge. They occur when experimentation is valuable. They're essential in creating something new such as a service or product. There are three types in a workplace."

- Preventable failure: a failure caused by deviating from a known process.
- Complex failure: a failure caused by a system breakdown.
- Intelligent failure: a failure caused by an unsuccessful trial.[8]

I'm a big believer in what can come from failure and believe we learn twice as much in failing as we do in success.

---

8   Amy C. Edmondson, Novartis Professor of leadership and management at Harvard Business School. Strategies for Learning from Failure, April 2011, https://hbr.org, 2011/04 strategies-for-learning-from-failure

Don't get me wrong; I'm not advocating large-scale failure or being irresponsible or negligent. Failure can be a horrible thing or a wonderful thing. There are catastrophic ones, and there are small ones. Some careers have an extremely low tolerance for failure, such as health care, the military, fire and rescue, policing, aeronautics, space travel, and mechanical. The space I'm addressing is in other environments where life is not on the line, and that's an ample space that includes some businesses, government, the nonprofit sector, education, and several others.

I want to share an example of failure with you. When I requested folks share some of their stories for this book, there was hesitation, a natural discomfort. That left me to bare my soul on failures during my time in food banking. There are so many to share, both big and small. I'll share the following because it's a failure in one of the most important areas (if not *the* most important)—leadership. Here goes the failure, what we learned, and what we did about it.

The food bank had just moved into our new one-hundred-thousand-square-foot facility, which included the Darden Community Kitchen, a state-of-the-art commercial kitchen designed for an innovative culinary training and job placement program. We had never done anything like this before, so perhaps it was inevitable that challenges were forthcoming, despite lots of due diligence in creating the program. When we began interviewing candidates for the role of executive chef, one person stood out among the others. She was dynamic and creative, had authored cookbooks and done television shows, and more. We hired her, excited to see her in action.

What transpired was a train wreck. The unrest started quickly among the kitchen staff due to conflicting orders, lack of communication, and disagreements on training and menus, all of which went unresolved. We soon realized that her standards did not fit our goals; they were unrealistically high and inappropriate for our nonprofit environment. On top of that, food costs were not controlled. While meeting with this chef to discuss these issues, her attitude indicated that she saw herself above everyone else. The team was starting to collapse. The time came when we released her.

So, here's some learning that came out of this. It was a failure on two levels. First, in recruiting the chef, we were naive and did not have an understanding of kitchen operations and financial management; we assumed all of this and fell for the personality. Second, it was a failure of leadership on the chef's part. She turned out to be only interested in herself, lacking critical leadership qualities. We learned a valuable lesson, and in our conversations about hiring another person, we discovered we did not need an executive chef;[9] we needed a food service manager that other chefs complemented through a combination of volunteers and paid positions. The result of the new hire was an incredible turnaround.

Not only did we get the food service professional, but also a phenomenal leader, Nancy Brumbaugh. She has created a strong and highly functioning team and expanded our original concept a hundredfold into catering. She makes a few signature products sold at retail and carries out high-volume meal production. The kitchen, now expanded with an added facility, produces approximately fourteen thousand meals daily and delivers hot and cold meals. And all of this was done within a budget with no surprises. Nancy has created a family-like environment.

The word "failure" often strikes fear in many organizations as something to be avoided. However, avoiding failure and fearing it is not a good thing. It's part of reality, don't let its negative energy limit you. "Avoiding, fearing, and punishing failure can harm your company and employees. If you expect your employees to get things done right the first time, they experience more pressure and get discouraged from trying new things. That can stifle creativity and innovation."[10] Another element of failure is fear, one of our most powerful emotions. Fear is ever-present when it comes to failure. My point is that if we let it, fear can stop us from experiencing

---

9   This is not to imply that executive chefs are not needed in any kitchen. They are vital to the success of a food establishment. Our program and nonprofit culture differ greatly from the for-profit model.

10   TandemHR.com. Why It's Advantageous To Allow Failure Into Your Workplace, February 24, 2022, https://tandemhr.com/why-its-advantageous-to-allow-failure-into-your-workplace/

so much in our lives, careers, and dreams. Anxiety in the workplace is often paralyzing and prohibits advancement and learning. Failure can help propel you to see your situation from different perspectives and find approaches to succeed that you might not otherwise have found.

Failure is often connected to finding fault. Someone is expected to take the blame, and the failure is typically buried and unaddressed. We must not play the blame game. According to the Harvard Business Review, "Develop a clear understanding of what happened—not of who did it—when things go wrong."[II] When we fail, we merely join the grand parade of humanity that has walked ahead of us and will follow us. It hurts, but with the right attitude, we can move on. Accepting failure is critical, but we must close it off and avoid ruminating. In the words of Debbie Macomber, "It's all right to sit on your pity pot now and again. Just be sure to flush when you're done."

As a leader, I always believed it was essential to create a safe workplace around accepting failure; we can learn so much from it. The safe space can create a culture that diminishes the blame game. I always focused on what happened versus "who did it?" We must recognize that failure is inevitable, change moves quickly, and organizations are often complex. I have found that failure is rarely caused by a single person or factor, but by a few that add up. The alternative of putting our heads in the sand is not acceptable. I love the following quote from Tennessee Williams; it takes such a realistic and inspirational attitude: "You will be wrong, and you will be bad quite often. That is the process of growing. Keep failing, but keep listening and keep learning. Do not let the failures allow you to shrink and move into some small corner to do your work. Always be big and bold. Take risks. No one grows without much stumbling."

It's necessary to build a culture of accepting failure. Once this is accomplished, staff are more likely to share openly. As a

---

[II]   Amy C. Edmondson, Novartis Professor of leadership and management at Harvard Business School. Strategies for Learning from Failure, April 2011, https://hbr.org, 2011/04 strategies-for-learning-from-failure leader, you may have to be upfront with your team and share a personal loss as an example. This kind of transparency, when handled properly, is an excellent example of behavior to model.

This next statement may sound counterintuitive; however, I believe it rings true. "In our lives, and in our organizations, most of us would benefit from experiencing more failures, not fewer." [12] I believe mindset is so important; embracing failure is part of leadership. Be out front with it, don't hide it. Here's a story from my food banking days that illustrates this belief. Five or six of us CEOs from various food banks decided we would visit each other's food banks with some of our board members. We would have dinner, then spend the next day sharing our latest initiatives and what we were most excited about and doing a tour. When it came time for me to host the group in Orlando, we had already been to four other food banks. I wanted to make the most of the tour because, by this time, they had been on several. After all, how many coolers and freezers can one walk into? So, I conducted a "failure tour"! I had my senior leadership team stationed at various locations throughout the food bank, and they shared a failure, what we learned, and what was done about it. Examples of operations, programs, and fundraising were shared. At the end of the tour, one of the CEOs pulled me aside and said, "I am so glad you talked about X, Y, and Z because we lost our shirts on those, too, and didn't deal with it correctly."

For those of you who attend conferences, you know that when people share what's going on with their organizations, it's mostly a brag session. Everyone likes to talk about how great things are going. I'm not saying there's no benefit to sharing best practices; however, let's not deny that we're imperfect. Some of the most valuable conversations I've had were those in which people were transparent and willing to be open about mistakes and the learning that resulted. When I was visiting other food banks that had built new facilities, in preparation for designing our own in Orlando, one of the most valuable questions I asked was, "What would you do

12   *"The big idea: why we need to learn to fail better"*, Amy Edmundson, The Guardian, August 28, 2023

differently if you did this again?" Failure avoidance reduces mistakes, assumptions, and oversights.

Let's change our attitudes toward failure. If you don't fail, you don't learn. Are you pushing the limits of what could be done? When trying to invent the lightbulb, Thomas Edison said, "I have not failed 1,000 times. I have not failed once. I have succeeded in proving that those 1,000 ways will not work." Babe Ruth, Sultan of Swat, had 714 home runs in his career—along with 1,330 strikeouts, considered abysmal during his time. He redefined failure in baseball. He said, "Every strike brings me closer to a home run." There are countless examples like these over time.

A fun and creative way to acknowledge failure comes from a group of entrepreneurs. They hold occasional "F—— Up Nights"; they tell their stories about loss, followed by a Q&A session. Everyone takes a turn at the mic. Sign me up! Some companies hold an "idea funeral." People gather and take turns eulogizing the failure they will bury, focusing on its merits and why it failed. I always thought having a national failure conference would be an excellent idea; what could come from that?

By shedding some light on failure in the workplace, some of the taboos and misunderstandings can be understood, and as a result, you can release more energy into your team and mission.

Keep It Light

Before looking at organizational culture, I want to mention two other leadership factors that are rarely addressed in the literature: anxiety and lightness. Anxiety is just below the surface for all leaders. This aspect relates primarily to the times we are living in. Anxiety is caused by our fast-moving era, a demand for instant results, challenging workplace situations, struggles balancing work and life, current events like fake news and wars, pressure to meet goals, and so on. As Fr. Richard Rohr, a Franciscan monk, said:

> Our age has been called the age of anxiety, which is probably a good description of this time. We no longer know where our foundations are. When we're not sure

what is certain, we will be anxious when the world and our worldview keep being redefined every few months. We want to get rid of that anxiety as quickly as we can. I know, I do. Yet, to be good leaders of anything today, we must patiently hold a certain degree of anxiety. The higher the level of leadership someone has, the more anxiety they must be capable of holding. Leaders who cannot hold anxiety will never lead us any place new.[13]

Future leaders will be significantly challenged to master anxiety as our world becomes more complex.

The second factor I mentioned is lightness. We are surrounded by challenges, issues, problems to be solved, and more; it can and does get heavy sometimes. So, if we lighten up, it may well light up those around us. At weekly senior leadership team meetings, I used to build in an informal check-in from everyone. These check-ins could include a funny story with a volunteer, coworker, donor, or family. It could be about a movie they saw or a book they read. It might also include poking fun at each other or us. I found that once one person shares, it opens the door for others to do so. Some of our biggest laughs happened during these moments. Over time, it set the stage for a more comfortable, less formal space and many smiles.

Throughout my interviews, a variety of qualities of good leaders were shared, some of which appear below. These traits are universally valuable in nonprofit organizations and for-profit entities.

- Be a good listener. Remember the adage "You have two ears and one mouth."
- Be a facilitator. Know how to lead conversations, not dominate them.
- Be a storyteller. We all learn from stories.
- Be transparent and vulnerable. Sometimes these are considered weaknesses. However, they are strengths when boundaries are set.

---

13   "Richard Rohr —Anxiety, Ambiguity and Identity," What Do We Do With Our Fear?", Center for Action and Contemplation, Fr. Richard Rohr, March 29, 2022, https://cac.org/daily-meditations/what-do-we-do-with-our-fear-2022-03-29/

- Be honest. Integrity is important. Nothing will take you down more quickly than a lack of these traits.
- Be compassionate for people and mission first and foremost.
- Be a communicator. Learn to speak different "languages." Your relationships may include businesses, elected officials, the faith community, philanthropists, volunteers, people needing your services, and many more.
- Be adaptable. If you aren't flexible, you'll snap.
- Be a servant leader. It's not about you; it's about them.
- Be solutions-oriented.
- Be able to deal with complexity. The world is not all black and white. Be aware of all the dimensions of an issue or opportunity.
- Be a bridge spanner. Look for connections among partners in your community.
- Be forward-thinking. No matter how successful your past was, your most important goals are in the future.
- Be willing to learn. Watch out for your blind spots.
- Be curious.

Culture

The second major requirement for an organization to succeed is culture. As Fons Trompenaars and Charles Hampton-Turner wrote, "A fish only discovers its need for water when it is no longer in it. Our own culture is like water to a fish. We live and breathe through it."[14] Culture is the water we swim in, or, as my friend Kyle Johnson, CEO of Central Florida Lighthouse, says, "culture is the oxygen you breathe." Culture can help or hurt an organization's ability to grow and achieve its mission. Taking time to make culture visible and improve upon it is critical to organizational success. Culture is much deeper than "lunch and learns" and birthday cakes. It's about attitudes, beliefs, mindsets, and traditions that answer one question: what does a company or organization value most?"

[14] Fons Trompenaars and Charles Hampden-Turner, *Riding the Waves of Culture, Fourth Edition: Understanding Diversity in Business* (McGraw Hill, 2020), Chapter 3.

Culture refers to the set of values, ethics, and beliefs that define day-to-day operations and the atmosphere at an organization. As Sujan Patel wrote, "Companies with positive, vibrant cultures attract the best talent to their teams and produce better products and services since their employees are more motivated to give 100 percent daily."[15]

I was once discussing culture with my friend of thirty-plus years and founder of the Atlanta Community Food Bank, Bill Bolling. His question was, "Why is it a Publix or Kroger that makes you feel good and smile when you walk into any of their grocery stores? What about Chick-Fil-A? They all draw from the same labor pool. Recruiting and training must be important to the CEO and the front end. Perhaps their employees are more empowered?"

I recalled a Publix store I frequented about once a month, where I went one day to pick up a prescription. I was checking out at the cashier, and as I turned to go to the pharmacy about thirty feet away, the pharmacist looked at me, smiled, and said, "Mr. Krepcho, I have your prescription ready for checkout." I was so impressed that she remembered me, called me by name, anticipated me picking up the prescription, and had it ready in a bag. I am not used to that kind of attention. Whenever possible, my wife and I will shop at Publix. I don't care if we drive a few more miles to get there. Whenever an organization or company can make you feel good during an interaction, the door opens to a more positive connection.

Culture is the foundation on which you, as a leader, build your organization. Once that is successfully in place, all the other activities necessary to operate can become even more impactful. When things are aligned, more progress is made. Goals, strategies, talent, budgets, and plans are all critical; however, excellent results can only happen when placed on a healthy culture's foundation. A higher level of energy can be harnessed. Culture is the "secret sauce" of success. Within a culture, you build your team. Author Margaret Wheatley

---

15   Sujan Patel, "10 Excellent Company Culture Examples for Inspiration," Entrepreneur, April 26, 2023, https://www.entre-preneur.com/growing-a-business/10-examples-of-compa-nies-with-fantastic-cultures/249174.

wisely states, "In this relational world, it is foolish to think we can define anyone solely in terms of isolated tasks and accountabilities. We need to conceptualize the energy flow patterns required for that person to do the job. We must see any person's role as where energies meet to make something happen."[16] A peer in the food bank network said, "We don't get many chances to work with like-minded people who are dedicated to the same task; it is an honor and a privilege. Culture is a verb." Let's make the most of the privilege for everyone's sake.

There are numerous examples of how to reinforce cultures, some very simple and some complex and detailed. I'll share one tangible example of something that was done at the food bank I led. Volunteers are vital to the food bank's mission and an essential part of its culture. They receive a sticker that identifies them when they enter the food bank; it's pretty basic, but when we get large groups coming in, this helps everyone know who the volunteers, other visitors, and staff are. The sticker is a simple badge of honor as well. When volunteers return for future visits, they are given a sticker labeled "Returning Volunteer." This acknowledges their commitment and allows staff to thank them for returning, as they may pass each other in a hallway, break room, or distribution center. Similarly, I was aware that people can get a feel for a place as soon as they walk in. I took this to heart and claimed that the food bank did not have a "receptionist," but we did have a "director of first impressions." Every person who was greeted and every phone call that was answered made the first impression.

Over the years, I have been surprised to see how few organizations examine their culture and realize its effect on their impact in either a negative or a positive way. Staff needs to be connected to the fundamental identity of the organization. Who are we? Who do we aspire to become? How will we plan to interact? Culture is the foundation for all else. It exists whether known or not. When the culture is right, all

---

16   Margaret J. Wheatley, *Leadership and the New Science: Discovering Order in a Chaotic World* (Berrett-Koehler, 2006), 72.

else flows from it. As Bahiyyah Maroon, an urban research institute's cultural anthropologist, said in an interview with me, "Culture goes beyond measuring work output; scales of reciprocity determine organizational culture. How can we help each other? Is it balanced?"

In a workplace, company culture involves the entire team; it connects everyone in a shared understanding. While I was leading Second Harvest Food Bank of Central Florida, folks visiting from the outside would often comment positively on our unique environment. We realized we had to determine what was unique and be more intentional about our culture. We identified our culture through a comprehensive process that included every staff member, from truck drivers to upper management.

Being inclusive of all staff and gathering their input was worth the effort. We also wanted to combat any feeling of separation between "carpet" and "concrete" people. While the leader of an organization must be the "culture keeper" and walk it daily, all staff must be included in identifying it. We sought input by asking small focus groups about various qualities of our workplace and what made it special. We used the Appreciative Inquiry method—a collective inquiry into the best of what is, that also imagines what could be—followed by a collaborative design of a desired future state. [17] An added bonus of the process was the incredible stories we heard about what brought people to the food bank's mission. A deeper understanding and respect for each other surfaced.

After the focus groups, we synthesized the input to define our organizational culture. Then we shared it again in two additional sessions to further distill it into a statement. Once established, this statement was prominently displayed, signed off on, reinforced in different ways, used in job interviews and new staff orientations, and valued in performance reviews.

Members of every organization must reflect in their own words on what's most important to them. There is no

---

[17]   David L. Cooperrider and Suresh Srivastva, "Appreciative Inquiry in Organizational Life," in *Research in Organizational Change and Development, Vol. I,* eds. William A. Pasmore et al. (JAI Press, 1987), 129–169.

one-size-fits-all solution to shape culture. Second Harvest
Food Bank of Central Florida's culture statement is an example.
Second Harvest Food Bank of Central Florida:

- We take pride in being part of a team.
- We treat each other like family.
- We respect each other and listen to all perspectives.
- We trust each other.
- We value transparency.
- We innovate today to create our tomorrow.
- We proactively embrace the concept of growth for all.
- We honor our commitments.

I want to share a touching story, told during one of the
focus groups, that is an excellent by-product of the whole
process of articulating culture. Moreover, it speaks to valuing
transparency, which is part of our statement. A hard-working
man named Josmy had worked at the food bank for eight years
in the distribution center, quietly going about his work. I think
part of his Haitian culture may have kept him more private.
He performed several laborious jobs without complaint.

When we got into our focus groups, Josmy spoke up about
why he likes working at the food bank. In his response, he
told us of his upbringing in Haiti in a large family, very poor.
One day, his mother came to him and said, "You're going to
school." He discovered that a missionary woman had offered
to pay for school for one of the children, and he had been chosen. None of his siblings had the opportunity. He graduated,
moved to Orlando, and was hired by the food bank. He told
the group that not a morning goes by that he doesn't think
of the kindness of a stranger who dramatically changed his
trajectory in life. In thanks, he brings dedication to his role
and supports family members who still live in Haiti. Josmy has
since attained further promotions and even greater responsibilities at the food bank.

It would be great to say that we had it all figured out and
that the initiative was completed. But maintaining a healthy
culture takes vigilance, as much as we would like to think it's
on autopilot. I want to share a case that has certainly affected

many organizations and will continue to do so in the future as the world workplace transitions into a more virtual one. We hired a few dozen new staff during COVID-19, and they worked virtually. Some of us were talking about our culture one day, and one of the virtual workers looked dismayed and said, "What culture?"

What a wake-up call she provided. How do you keep a culture when people are working remotely and do not know the rest of the team or have a feel for the environment? One activity we did at the food bank during COVID-19 that helped in a small way was a voluntary half-hour virtual lunch get-together hosted by HR. We called these "mission moments." They were thirty-minute Zoom sessions featuring one of the teams of the food bank. Topics ranged from what was happening in operations or programs, the financial security of the food bank, inspiring stories, and much more. Despite the challenges and devastation that COVID-19 presented, it also offered creative opportunities like this one that will live on into the future.

So, whether an organization is a nonprofit, foundation, business, government body, or place of worship, it has a culture. As leaders, we must identify ours, clearly state what it is, and then promote, practice, preserve, and celebrate it.

Vision

Once the first two legs of the stool—leadership and culture—are in place, what secures the organization's strength is the vision statement. For me, establishing a vision for an organization is one of the most exciting things leaders do because everything flows from that statement. The leader's critical decisions and direction, allocation of resources, communications, and day-to-day practice must always go back to the vision. Are we working toward it? Does our budget reflect the image? What about programs? Organizational environment? Language?

So, what exactly is a vision? I define it as a framework for possibility. The idea must be significant, aspirational, and have a vast reach. It must also be engaging, full of potential, exciting, and hopeful. A friend and colleague said people mistake vision-setting for some magical way of seeing the future. Instead, it comes from a deep understanding of the environment that you're working within. It's an opportunity to see elements that have not been there before. They can be arranged for new possibilities; it's a way of seeing what's not realized yet. It must go beyond the status quo.

Over the years, I have been surprised by how many nonprofit organizations still lack a vision. To run a nonprofit organization without a clear picture will severely limit the potential community impact. As a leader, whatever vision you decide on, pay attention to what you're declaring. I want to explain what I mean by "declaring." I have seen over the years that organizations claim to solve their issue within a specific time frame, for example, "End world hunger by such-and-such a date." If this is your mission, you're setting yourself up for failure because of unrealistic deadline expectations. When you fail to reach your goal in that time, you and your staff will be disappointed, as will the rest of your audience.

The second point I want to make is a much larger one. In the food security arena, food bankers use two different phrases to describe their vision: "end hunger" or "fight hunger." Do you believe hunger can be ended? Or do you think it cannot, and you'll fight it forever? What you choose has enormous ramifications for your goals and strategies. Depending on the desired vision, two organizations in the same space will look very different.

Here's a modern-day parable that illustrates the vast difference between several approaches to solving serious issues. It clearly shows that perspective on an issue will dictate the solution an organization chooses. In this tale, two people are walking by a river when they notice babies floating down the river. One grabs the babies from the river while the other runs upstream. The first asks, "Where are you going? We must save these babies," and the other replies, "I'm going to see who's throwing babies in the river." Eventually, the people figure out the source of the problem rather than just dealing with the symptoms, and no more babies are endangered.

To solve the problem, you must go upstream and find the root cause. For food banks, I don't believe it's an either-or choice; I think it's a "yes, and." Changing systemic issues in the best of times will take years, so we cannot stop feeding people today. We must develop the bandwidth to address the more significant systemic issues that require new mindsets, as well as talent, resources, partners, persistence, public policy, and patience. We must think deeply, alongside staff, board, and community, about what business we're in.

My friend Brian Greene, CEO of the Houston Food Bank, said, "Food security isn't about food; it's about income." This perspective has significant implications for Brian's mission, goals, strategies, programs, partners, and messaging to the community. Brian's statement starts to question the issues related to hunger, such as minimum wage, education, lack of affordable housing, and various other factors. Another longtime friend, Matt Habash of the Mid-Ohio Collective, said, "Hunger is a public health issue." Matt's perspective emphasizes a more specific priority that will drive various essential decisions. Matt's statement leads him toward healthy foods and nutrition. When we change the lenses we're looking through and focus on our vision, our future paths can start to look very different. This is not to say that one is wrong or right; they are just different in focus. Currently, there is a strong movement in the Feeding America network to place greater focus on ending hunger. This is yet another milestone in its history.

Here are two examples of vision statements that demonstrate the direction and perspective of an organization. The Feeding America vision is "An America where no one is hungry." The words "where no one is hungry" carry so much weight and are aspirational. When a vision is created, leaders must be prepared to own it and live up to it. At Second Harvest Food Bank of Central Florida, the vision is "To inspire and engage the community to end hunger." Unpacking that vision is what captures the hearts and minds of followers. By accomplishing this, I believe the resources for your mission will follow you.

As with defining culture, an organization's vision should be established with the input of multiple stakeholders. The board, the staff, the people who receive its service, donors, and others can provide valuable insights to help a leader craft their statement. It is not the task of the CEO alone to create the vision. The CEO will obviously have a vested interest in the direction and must lead and convene the discussion and eventually drive it. As a leader, cast the vision and let it belong to everybody because you will only achieve it through ownership.

The vision should permeate everything your organization does. Vision must be embedded into strategy, programs, budget, advocacy, messaging, and other areas. At Second Harvest, a further step was taken when designing the future Orlando food bank in the early 2000s. We asked ourselves, if we were to inspire and engage the community, what should the food bank look like? Feel like? What would reinforce our vision? What would make a visitor want to be involved in the mission? We incorporated vision into architecture, workspaces, and design. This may sound counterintuitive, but our vision was about people first, then food, rather than vice versa. Engaging people, whether as volunteers, donors, staff, or neighbors, makes the vision possible. It's not about us but about them.

We gathered the project manager, architects, interior designers, volunteers, board, staff, and our partner feeding programs to be included in the design process. They provided ideas that were implemented and live on today. As one example, consider the main entrance lobby we designed. It's a two-story atrium, with windows that flood it with natural light and a twenty-four-foot glass wall that looks directly into our operations. Colorful graphics, images of people being fed, a mural reflecting our mission, and an acknowledgment to community partners immediately grab visitors' attention.

We placed colorful murals, inspiring statements, and video monitors throughout the facility, and we displayed live tracking of our operation's key performance indicators for visitors. We thought long and hard and decided not to spend too much money and instead to strike a balance between stewardship and an inviting and credible environment. Our special friends at Walt Disney World then took the vibe to a whole other level with murals and creativity throughout the food bank's public spaces. It was impossible to walk through the facility without a big smile. A few locations became photo stops! Visitors wanted their pictures taken in front of various murals. The space conveyed a solutions-oriented vision rather than one depicting suffering people with swollen bellies. Not only were visitors engaged, but they shared their experiences with others on social media.

Another building-design element was an event space accommodating up to nearly two hundred people. By making this room available for community meetings, parties, weddings, bar and bat mitzvahs, and other special events, over twenty thousand people visited the food bank in the first four to five years. These folks were exposed to our mission up close during their visits and tours. Most of those visitors had not visited the food bank before the creation of this space. As a result, more people became engaged and inspired, helping to fulfill our vision. The vision for the future facility was incredible in galvanizing the board and staff because of the belief and hope for the future it represented. Not a day went by without a visitor coming to the food bank, walking into the lobby, and saying in amazement, "I had no idea this is what you look like!" And after a tour, we often heard the affirmation, "You're much more than a food bank!" A well-thought-out vision can undoubtedly translate into greater impact and success.

I have described the three legs of the stool: leadership, culture, and vision. Imagine if one of those legs were two inches short. We have all been to a restaurant at one time or another where just a fraction of an inch of space unbalanced the table, which was annoying. We always found something like a matchbook or napkin to place underneath and make it stable. Inevitably, that solution didn't last long. Here's to stable tables and sturdy stools.

Chapter 12
Conclusion

The best way to predict the future is to create it.
— Peter Drucker

When I originally drafted this chapter, I wanted to answer the question: What does the future of food banking look like? I have no crystal ball. Environments change, and no one can predict the future. Who knows what influence population growth, climate change, technology, politics, war, the economy, or social inequalities will have in future times? It's the chaos we live in. After much introspection, I realized the answer depends on another key question: Do we want to *fight* hunger, or do we want to *end* hunger?

Depending on your answer, very different approaches, strategies, designs, and decisions will be made. If you decide to fight hunger, God bless you; what an incredible impact you will continue to have on countless souls. Go forth and do more of it, by all means.

After my thirty-four years in food banking, I have concluded that we must feed people while simultaneously working to end hunger. *We will not food bank our way to ending hunger.* This thought has been shared in some circles and may sound harsh to some in food banking, but it needs to be said. I am not saying that food banks are a failure and should shut their doors. They are impactful, much needed, and will continue to help improve the lives of millions. I could not have spent over three decades in food banking if I didn't believe it had incredible value for neighbors and communities. I believe in the people who are dedicated to this calling. However, providing greater quantities of charitable food is not the single solution. Mark Brewer, president and CEO of the Central Florida Foundation, rightly states, "One of the issues I've seen when a community is dealing with anything in the public space is that much of what we're trying to deal with are symptoms of other problems. These are big, complex social issues that there's no single switch to flip to solve. You need to flip many of them and some in parallel, and there can be many unintended consequences."

We can end hunger; it is possible. It's a damned intractable problem; it can be solved, but not in the current model. I'll use an example to illustrate the enormity of the challenge. Let's say that to feed every person facing hunger for a year, it would take three times the amount of food that Feeding America and others distribute.[1] That would amount to approximately sixteen billion meals. If you lined up that many place settings side by side, it would circle the globe 121 times or go to the moon and back six times. Even if that amount of surplus food could be secured, at considerable expense, the same (or more) food will be required the next year and the one after that. How is an effort of that size sustainable? To make matters worse, the "solution" does very little to address the reasons people are facing hunger in the first place. The source problem would remain and become even more significant with a growing population.

Those of us in food banking must keep the food bank doors open but reframe our approach. Many of us now realize that it will take more than food. There are so many factors at play that impact food insecurity. Brian Greene, CEO of the Houston Food Bank, says, "Food insecurity isn't about food; it's about income. Income security is manifesting itself so often as food insecurity. When we provide food to a family, we are also helping them pay rent, utilities, or other expenses." Brian continues, "In the early days in the Feeding America

---

[1]   This is only an estimate, for illustrative purposes. Distribution in 2022 was 5.2 billion meals.

network, the prevailing thought was that if we just distributed enough food, we would end hunger in America. We were wrong. The network long ago reached the amount of distribution that would have theoretically eliminated the meal gap of the 1980s. Yet there is just as much hunger in the U.S. as when we started."

To illustrate what Brian said, I recall an editorial I wrote about a month after I started as CEO in Orlando in 2004. I was shocked at the number of hungry children, approximately 95,000, in a six-county area of Central Florida. In the last editorial I wrote just before I retired seventeen years later, in 2021, I stated there were now 157,000 children facing hunger in Central Florida. This was after providing nearly enough food for six hundred million meals in Central Florida over that period.

The line is becoming longer for people needing support. The poverty rate hardly budged over those years. Nearly half the people receiving the food were children; most of the rest of it went to working households, seniors, and the disabled. A small percentage went to the homeless population. So, what's wrong? The economic system is f——d up across this country for millions of middle-income and, predominantly, low-income households. As mentioned, many community issues have gone unresolved, and some have worsened over time, such as the lack of affordable housing, decent wages and benefits, access to health care, and dependable public transportation. Meanwhile, many people with healthy incomes are doing just fine and separating further from the pack. I am not against capitalism and believe wealth is incredible; the food bank could not have responded so well over the years without generous people contributing. It's about the dramatic inequality that exists. The line of people needing support continues to get longer.

I believe the immediate future approach for food banks must be a "yes, and" goal, meaning we must feed the line of people today while at the same time working on the deeper issues and thus shorten the line of people needing help. The core competency of providing food must not be lost, but there is an even greater opportunity to join others in more comprehensive solutions to hunger. The future includes partnerships that did not exist before. We have to look at the bigger picture beyond our own singular view. Food banks won't shorten the line of people needing help by themselves; they will have to partner and collaborate even more so with others.

Food banks and nonprofits vary dramatically in size, resources, and philosophies. Less-resourced organizations may look at the "yes, and" approach with dismay because they have their hands full with the day-to-day challenges. I acknowledge that reality. It will take all of us to end hunger in our own unique and measured ways. It will look different in every community. Kudos to a few food banks that are intentionally reevaluating their budgets regarding what percentage goes to fighting hunger versus ending hunger.

The "yes, and" approach for food banks is still a relatively new trend. If wider adoption is realized, this could be a historical inflection point for the food bank network. Several food banks have started, all at different levels, to adopt the practice. Most, however, have yet to embrace this direction. I view this approach as "Food Banking 2.0." Ten years later, it will be recognized as another key milestone for the Feeding America network.

I recently read about "liminal space," a state where we can begin to think and act in new ways. It is where we are betwixt and between, in transition, having left one stage of life and preparing to enter the next. Food banks are betwixt and between in how they look at fighting and ending hunger. Fr. Richard Rohr, a Franciscan priest, speaks of liminal spaces in a spiritual sense. I can see how it also applies to food banking, which I believe is a spiritual experience. He states, "Liminality is a form of holding the tension between one space and another. It is in these transitional moments that authentic transformation can happen. Otherwise, it is just business as usual and boring, status quo business."[2]

So, what exactly is this "yes, and" approach? How do we feed the line of people today while attempting to shorten the line of people for tomorrow? So much great work has

---

2   Vanessa Guerin, ed. *Oneing: Liminal Space* (Center for Action and Contemplation, 2020), 17.

started among nonprofits. For their part, food banks across the country are starting to approach this challenge in various ways. In 2020, upon completing the food bank strategic plan in Central Florida, four critical upstream issues that impacted hunger were identified. Although there are more, the following were prioritized based on the food bank's capacity as an organization: health, education, income, and housing. We were already doing work on three of these areas, but realized we could do much more. The food bank had not previously looked through that type of lens with such intention. The housing issue was a stretch, but we wanted to explore opportunities to collaborate with the affordable housing sector by complementing their efforts, not duplicating them, or starting mission drift. The next logical step would be to examine these four areas and see them as part of a system rather than individually.

The "yes, and" approach should be done incrementally; otherwise, staff and financial resources can be strained. Depending on its capabilities, a food bank can start with one upstream issue or multiple. Some of these upstream issues have a steep learning curve, so solutions can take time to be effective. We need to do more systems thinking around the issues relating to low income. According to Veronica Hotton, systems thinking is:

> A way of conceptualizing and understanding how various system elements are related to and influence one another. It could be a company, a family, a supply chain, or a country. Many of us have heard the metaphor of an iceberg: ten percent above water; ninety percent below the surface—being impacted by all kinds of things we can't see, like ocean currents, sea life, and other factors. What is seen and unseen? Diving below the surface, we can see patterns, structures, thoughts, and larger paradigms at play.[3]

3   Veronica Hotton, "The Iceberg Model of Systems Thinking,"
       Dumbo Feather, March 24, 2022
       https://www.everand.com/articla/568400603/
       The-Iceberg-Model-Of-Systems-Thinking

At the base of the iceberg are the deep-rooted factors: our beliefs, attitudes, theories, morals, expectations, and values, which allow structures to continue functioning. If we ignore the bigger picture and look only from an isolated viewpoint, out of context, we can't possibly solve the thorny issues. We must become part of a more extensive system.

Systems thinking has begun to show promise in the tri-sector business model. Significant community issues will not be solved with only one or two entities working alone. Successful models and programs must be scaled up, and this can only be achieved through increased community participation. In the tri-sector model, the private, public, and nonprofit sectors sit at the same table, and each contributes to and benefits from the model for the good of the community. The table has seats for people from the impacted groups: business, government, education, health care, transportation, nonprofits, nongovernmental agencies, faith groups, and more. And when sitting at the table, personal agendas and egos must be set aside.

This is challenging work to do. At the core of any success in this systems approach is trust. Care must be taken that a top-down solution is not handed down from above. It must be bottom up. A growing number of programs are based on the tri-sector model. An example is Propel, an organization that builds technology designed to help low-income people improve their financial health. It utilizes a tri-sector approach to enhance the multibillion-dollar SNAP program. The solution enables app users to track their account cash flow, access discounts, download coupons, and find resources such as food pantries, farmers' markets, and other social services. In addition, it leverages the USDA's investments in SNAP.[4]

Food banks can work within this tri-sector model (some already are). By working with other nonprofits, local government, and the business sector, I believe food banks have an opportunity to create "hunger-free zones." The systemic approach is the only way to achieve that vision. There are

---

4   "Tri-Sector Business Models," New Impact, 2023, https://www.newimpact.care/tri-sector-business-modelsı.

institutes across America dedicated to various topics; why not hunger and poverty at the community level? I could see a local college or university being the host, bringing academic rigor, structure, and varied voices to discuss the issues and solutions. What about starting communities of practice around the issue, or innovation accelerators? Can we find models elsewhere that work? Bring them back, pilot one, and tweak it for our local community? Building and maintaining hunger-free zones would require long-term commitments, support in data and technology, advocacy, policy, diversity, equity, inclusion, and additional ways of reducing food waste.

It's not a matter of having enough money; it's a matter of priorities. Mark mentioned that giving as a share of GDP has rarely strayed far from 2 percent over the past four decades; imagine if it could increase to 4 percent. I saw an interesting graphic that showed a cup of coffee with the statement, "If everyone gave up a morning cup of coffee, $220 billion more could go to charity."[5] No one will be given a free ride when all these elements come together. The opportunity exists, and communities will benefit as new possibilities are presented and barriers broken down. Let's channel this country's incredible riches with imagination and hope. We have choices. Continue with the status quo, and the problem will become more significant —unacceptable and unsustainable. Our other option is to think bigger. Is there courage in the community to do so?

So, hypothetically, if effective and impactful collaborative work is done in a community, and hunger-free zones are achieved, what is the future of food banks? Food banks could then put an even greater focus on providing food to those confronting emergencies such as fires, floods, storms, and situational poverty. These are shorter-term needs, not chronic problems. Food banks would have more resources to leverage for this smaller group of people. They could then focus on shortening the line of people needing food. A few food banks and related organizations currently operate as community

---

5   Suzanne Perry, "The Stubborn 2% Giving Rate," *The Chronicle of Philanthropy*, June 17, 2013, https://www.philanthropy.com/article/the-stubborn-2-giving-rate/.

resource centers that offer multiple services, such as support with job searches, literacy, and health care. These models could be expanded with the resources freed from ongoing chronic food distribution. Could this be "Food Banking Version 3.0"? I believe we're at a threshold moment in the goal to end hunger; it's within our grasp. Let's envision a world that is better than we ever imagined and filled with hope.

My closing message for the readers is that all of you can and do make a difference: we can end hunger. It's within our grasp. Nonprofits need you to donate your time, talent, and treasure. Please advocate, donate, volunteer, and create hope for the future.

Epilogue
Humor and Funny Stories in Food Banking

There is nothing in the world so irresistibly
contagious as laughter and good-humour.
- Charles Dickens, A Christmas Carol

How can a book about hunger possibly have humor in it? The truth is, there's so much material to work with that I contemplated writing the entire book about comedy in food banking. Who doesn't enjoy a good laugh? People working in an organization that battles daily for a social cause may need a good laugh more than others. The food bank I led was filled with belly laughs, cackles, chortles, guffaws, horselaughs, snickers, and giggles. John Cleese said, "Laughter connects you to people. It's almost impossible to maintain any kind of distance or any sense of social hierarchy when you're just howling with laughter. Laughter is a force for democracy."[1] World leaders should have a giant breakout session at Davos and tell jokes. Or, how about if Congress dedicated daily time to roasting each other?

I want to share a few funny stories and a list of strange food donations I've seen over the years. This chapter will read differently than the others because it contains short examples

---

[1]   "John Cleese Explores the Health Benefits of Laughter," Open Culture, April 29, 2015, https://www.openculture.com/2015/04/john-cleese-explores-the-health-benefits-of-laughter-yoga.html#google_vignette.

with no goal other than being a bit of entertainment. Please don't interpret my love of humor with an attitude that everything's a joke; it certainly is not. Yes, there are many times that require total seriousness throughout the day. Think of humor as a seasoning to the day; use it appropriately.

Kathleen DiChiara from New Jersey recalls getting a call years ago from the police saying they had just shot a steer that had escaped. "Can you take it to the food bank?" Kathleen called her husband during a business meeting and asked him what they should do with the bull. He didn't believe what he was hearing. They got heavy equipment to lift the steer and brought it to a slaughterhouse. The next challenge was where to find freezers to store the meat. A full-grown steer can include 24 roasts; 28 T-bones; 10 sirloins; 10 tip sirloins; 28 rib eyes; 12 round steaks; various quantities of short ribs, flank steaks, stew meat, and brisket; and 150 pounds of ground beef. Kathleen solved the storage problem and enlisted neighbors to open their freezers and each store a bit. No bull!

Pam Irvine from Salem, West Virginia, tells of the "disco turkeys" her food bank received. The wings were frozen in opposite directions, giving the impression that each turkey was dancing. She also recalls the donation of hundreds of thousands of cereal boxes. They were donated because a small red ball was inserted into the package to incentivize buying the product. Kids were being injured by putting the balls in their eye sockets. The donor agreed that if the food banks would open each box and remove the ball, they could have them. Volunteers in food banks across the country came to the rescue.

Jaynee Day from Nashville, Tennessee, shared a few gems. Once, one of her staff members came into her office visibly shocked and said the FBI was in the warehouse and wanted to talk to her. Jaynee had accepted a donation of products containing pseudoephedrine, a drug that helps with allergies and nasal congestion. She thought her distribution sites could provide these to their food recipients and save them some money. Unfortunately, pseudoephedrine is also an ingredient in making meth. The FBI said her inventory was worth about $1 million. It cost her $5,000 to dispose of it properly.

Another time, Jaynee received a truckload of mayonnaise; unfortunately, she found out too late that it had expired some time ago and the packaging was faulty, so they had to throw it out. They loaded their truck to take it to the dump. A friend called a few hours later and told her to turn on the local news because they were discussing the "Mayonnaise Caper." Jaynee was baffled. She received a call from the mayor, who told her that just down the street from the food bank, a car had come around the corner, slid out of control, and hit another vehicle. He told Jaynee that they had traced the mayonnaise that had leaked from the back of the truck to her loading docks. Jaynee paid for the sand to cover the road, and the case was closed.

Jaynee also passed along a story from Caroline Lanier in Texas. One morning, Caroline arrived to open the food bank and saw a massive flock of buzzards circling overhead. Upon entering the warehouse, Caroline discovered thieves had cut a hole in her roof and attempted to steal dozens of cases of tuna fish. When the thieves were bringing the goods up by a rope, it broke, and the cans opened.

Sometimes food banks receive odd donations. Diane Letson of Chicago was offered live geese and an emu. Team members from Orlando, Florida, shared stories of being offered a live flock of chickens, fifteen thousand red noses, an entire MASH unit, and various articles of heavy winter clothing. Hot dogs were labeled in Hebrew; cereal boxes were printed in Russian. Food drive donations included spotted dick in a can from England. Chocolate-covered French fries and green ketchup. A truckload of tuna, delivered by a mermaid. Semitruck loads of hand sanitizer. Thousands of ornate flower arrangements from a massive trade show once made our receiving area look like a gangland mobster's funeral. A donation of "jelly" was made in Kentucky. It turned out to be a donation of KY Jelly.

In 2000, I was a food sourcer in the national office of Second Harvest. Each member of this team was responsible for developing business partnerships with national manufacturers and other companies in the food industry. In addition, we focused on specific donor categories such as Retail Grocery, Mass Merchandisers, and Non-Food.

During this time, Randy Starck of Feeding America was tasked with developing new sources of meat and poultry donations. Donations of protein were highly valued by our food banks but seldom received. The focus was concentrated on Tyson, Perdue, Smithfield, and other large processors. Over time, however, the food bank protein world expanded as food banks were matched to smaller processors, repackers, and distributors in their area. No donation source was too small. We welcomed sharing ideas, no matter how bizarre, and the office was introduced to a small farmer to inquire about "spent hens." We learned that when egg-laying hens aged, they were no longer deemed "productive" by the farmer. Since they were smaller than chickens raised for meat, it wasn't cost efficient for a large processor to further process them as anything other than animal feed. The phone conversation with Lester the farmer went like this:

Randy: Hello, I'm interested in your spent hens and how I can get them processed into chicken pieces.

Lester: (heavy Southern accent) Well, you'll need someone to wrangle the chickens into the cages. Stack the cages in the truck. If it's not your truck, you'll have to pay and have the interior hosed down. I know someone who'll do that. Then you're on your own.

Randy: Uh, I don't have a wrangler, don't have a truck, and need a company who can literally take the hen from "farm to fork." Do you know anyone?

Lester: Closest place to go is in Mississippi. You want a kill plant and further processing facility. I tried to do this for myself. It made no sense, so thought maybe I could donate the hens to the food bank and get one of those tax deductions.

Randy: Okay Lester, this is what I'm hearing. I need to hire a truck to pick up the hens. I need a wrangler to load the hens into cages, stack the cages, drive to Mississippi, pay someone to unload the live birds, pay someone else to clean the truck, pay the kill plant to

prepare the bird for further processing, pay the pro
cessing fee, and finally locate a food bank interested
in picking up a load of spent-hen chicken breasts at
the processing plant. Whoa, this is expensive! By the
way, how many hens are we talking about?

Lester: If the wrangler's any good, you might get three
hundred or four hundred birds in a truck. I'd guess each
bird would yield a couple of pounds of meat.

The office passed on the opportunity, and we learned
that to end up with eight hundred pounds of protein would
only cost us about $10,000!

Al Brislain from Reno, Nevada, has a phenomenal memory from the past forty years of his food banking career and
shared the following list of strange food donations. While
the standard reason for donating many products is related
to minor ingredient errors, products nearing the end of their
shelf life, or discontinued products, some of the reasons may
seem surprising. A classic donation was the raisin bran with
too many raisins— "three scoops instead of two." Several
dozen semitruck loads were donated. The company said they
did not want to raise the consumer's expectations, so, for a
short while, the food banks that received the loads had the
best raisin bran in town.

On another occasion, Feeding America was given more
than two hundred semitruck loads of Trix cereal when the
manufacturer determined that the red Trix was "pinker
than red." Although the issue related to the food coloring,
it never affected the taste. It was decided that consumers
might notice a difference and think they were receiving an
inferior product. Feeding America also received a donation
of about twenty semitruck loads of Golden Griddle maple
syrup. It was gifted because a tasting panel had voted three
to two that the syrup was not up to standards. What if one
of those panelists was a little hungover or had a bad cold?
More than $1 million was diverted due to one vote.

Other goods are donated because of their packaging.
Many products are tied to a movie, and movies occasionally
flop at the box office. The packaging has pictures of the

film, so the product is taken off the shelves. Additionally, food banks often receive the old, unimproved version when companies develop a "new and improved" product. On one occasion, Tide went to a super-concentrated formula and donated several semitruck loads of the old version. It saved many families in need of critical dollars on their budgets.

Sometimes manufacturers make errors in production or unintentionally create surpluses. A company that provided high-end entrées to restaurants donated over twenty semitruck loads of frozen trout stuffed with crab meat to a food bank. The processor had made a mistake, and the trout meat had a yellow tinge but no difference in taste. Another example from years ago was when McDonald's changed their McNuggets from a mixture of white and dark meat to all white meat, creating a massive surplus of chicken thighs and legs. The USDA stepped in to balance the market, and food banks received hundreds of semitruck loads of chicken.

Sometimes, even the food banks were unable to use a failed product. One example was Magic Shell. This was a topping that, when poured over ice cream, creates a hard shell that looks and tastes like a traditional dipped cone. The company developed a dozen flavors, including chocolate, butterscotch, strawberry, and cherry. The buyer rejected the product for some reason—dozens of semitruck loads. The manufacturer contacted the food banks and offered it as a donation. It was unpopular with the recipient food pantries and shelters and sat in inventory for months. The food banks inquired how long the product could last and were told, "You could put it in a time capsule."

Wrapples were another food product that could have done better. They were developed to simplify the preparation of caramel-covered apples. The product consisted of sheets of caramel separated by wax paper. The caramel would soften when slightly warmed and then could be molded around the apple. It did not sell well, and the company donated the product to food banks. Many food banks had difficulty distributing the Wrapples, and some tried alternative uses. The Atlanta Community Food Bank, working with Georgia Tech, developed the most creative alternative. The school was confident they could repurpose the Wrapples into fuel, but the company refused to allow the experiment to go forward.

I recall back in 2009 getting a forty-foot truckload of Gatorade donated, packed floor to ceiling and wall to wall. As I took a closer look at the product, I noticed Tiger Woods's face on the packaging. His scandal at his home had occurred the previous week; no retailer would sell that product. Kudos to Tiger for getting his life back in order. I also recall a national donation that was offered to me while working at Feeding America: gently used women's underwear. A company had tested this latest product and, for some reason, didn't let the women keep the lingerie. I soundly rejected the offer.

Margaret Linnane from Orlando shared a story from the 1980s. A company offered some delicious food and over a dozen cases of beer as a donation. She rejected it because of the beer, and the donor said, "You either take it all or nothing." She took the donation. At home that evening at about six thirty, she received a call from the police saying that the warehouse had been broken into and some beer stolen. It was probably more of a breakout than a break-in. Various people from the neighborhood would "visit" the food bank occasionally. Margaret received another call at nine that evening from the police; another break-in had occurred; she met the police and closed the shop again. At about eleven thirty that night, she got another call from the police about a break-in. Margaret met the police, locked up, and put a note on the door: "The beer is gone!" That was the end of the break-ins.

A critical note to all these crazy donations: food banks, to a large extent, cannot control what they will receive and are often surprised. Many loads are rejected when caught in time. Also, these types of donations represent a tiny fraction of the total products. Food banks continue to significantly advance in distributing healthy products, such as tens of millions of pounds of fresh produce.

Other moments of humor relate to events at food banks. Bill Bolling from Atlanta shared a story from years ago. Early in his days at the food bank, he received a large donation of damaged nonperishable goods, and the food had to be sorted;

not all was acceptable for redistribution. A busload of accomplished women from the Junior League came to volunteer, and during their shift, a tiny mouse ran out of one of the boxes. You would've thought that the screams could be heard miles away. That was the day the Junior League levitated.

The Feeding America network did studies in the 1980s, 1990s, and early 2000s to learn more about the people being served and those needing food. They involved face-to-face interviews that took fifteen to twenty minutes to complete. Former NBA All-Star and senator Bill Bradley volunteered at a food bank to participate. One of Bill's interviews was with a woman in her senior years. She welcomed him, and he sat down at a table; he introduced himself and proceeded with the questions. Much to his chagrin, he realized that by the time the interview was done, he had unconsciously snacked away and eaten a whole bowl of peanuts the woman had at the table. He was deeply embarrassed and apologized profusely. The woman looked at him and said, "Don't worry honey, ever since I lost my dentures, I just suck the chocolate off them."

One of my most awkward and embarrassing moments was when I delivered a twenty- to thirty-pound box of food to an elderly gentleman. He lived alone on the second floor of a building with no elevator and a rusty exterior stairwell. I had asked one of my staff members to put a box of food together as I rushed out the door. As I approached the screen door, the man welcomed me into his small apartment. Sitting on a couch next to his walker, he was pleased that someone got the word out that he needed food. I sat down and saw a bowl with a few un-popped kernels of popcorn remaining on the table. He mentioned that it was all he had to eat the last three days. He shared that he had no relatives in the area, and neighbors rarely saw him because his disability didn't allow him to go up and down the steps to this apartment often. He was primarily homebound. On one wall were a few photos and medals from the war in which he served, of which he was proud. I opened the box of food, and to my shock, on top of all the other items were packages of microwavable popcorn! I just wanted to disappear. He saw my look and could not stop himself from laughing. I'll never forget that experience.

I'll close this chapter with a Feeding America annual tradition that brought the network closer through a heavy dose of humor. The network was divided into three regions: Eastern, Central, and Western. After the closing dinner at the annual conference, the Central Region Association Players (CRAP) would give a performance. This group of food bank directors created and performed songs and skits rich in satire, poked fun at each other, and reenacted lightened versions of disagreements that had occurred during the year between food banks and the national office. The performance resembled the Saturday Night Live format. Some of their material was funny enough to be a hit on SNL!

Each performance had a theme, and I recall one based on opera, of all things. The showstopper was a song by Tina Osso, a good-natured director from the Midwest. When she came on stage dressed as a Viking wearing a horned helmet, sword, fur skins, and a metal-cupped, Madonna-like oversized brassiere, the audience rose to their feet. That is undoubtedly an image burned into the memory banks of many.

Appendix

Could You Survive in Poverty? Quiz

Could you survive in poverty?

COMPLETE THE QUIZ:
Put a check by each item you know how to do.

_____1. I know which churches and sections of town have the best rummage sales.

_____2. I know which rummage sales have "bag sales" and when.

_____3. I know which grocery stores' garbage bins can be accessed for thrown-away food.

_____4. I know how to get someone out of jail.

_____5. I know how to physically fight and defend myself physically.

_____6. I know how to get a gun, even if I have a police record.

_____7. I know how to keep my clothes from being stolen at the Laundromat.

_____8. I know what problems to look for in a used car.

_____9. I know how to live without a checking account.

_____10. I know how to live without electricity and a phone.

_____11. I know how to use a knife as scissors.

_____12. I can entertain a group of friends with my personality and my stories.

_____13. I know what to do when I don't have money to pay the bills.

_____14. I know how to move in half a day.

_____15. I know how to get and use food stamps or an electronic card for benefits.

_____16. I know where the free medical clinics are.

_____17. I am very good at trading and bartering.

_____18. I can get by without a car.

From Ruby K. Payne, A Framework for Understanding Poverty, 4th rev. ed. (aha Process, Inc.: 2005). Reprinted with permission.

Food Bank Timeline Overview

1967:   First food bank founded in Phoenix, Arizona, by John van Hengel.

1976:   The United States Department of Agriculture (USDA) and food industry provide support for a national organization and additional food banks. Tax Reform Act and Good Samaritan Act passed in California.

1977:   Eighteen food banks started.

1979:   Formation of a national network of food banks called Second Harvest. Board formed, and initial national food donors contribute.

1981:   US government creates program to redistribute surplus commodities, called the Temporary Emergency Food Assistance Program (TEFAP), and food is channeled through food banks. Jack Ramsey—first CEO of Second Harvest.

1983:   John van Hengel leaves Second Harvest and focuses internationally.

1984:   Second Harvest moves its headquarters from Phoenix to Chicago.

1989:   Kids Cafés become national (modeled on the original café in Savannah).

1991:   Second Harvest creates National Advisory Council with member food banks.
Food Bank State Associations start.

1992:   Second Harvest starts national Advocacy, Policy, and Government Relations office. First significant disaster relief effort (Hurricane Andrew).

1993:   Second Harvest conducts the first-ever national hunger study

1994:   Second Harvest creates the School Backpack Program, which goes national in 1996.

1996:   Congress passes the Bill Emerson Good Samaritan Law that encourages more corporate food bank donors.

1998:	Second Harvest expands the mobile pantries that started in Grand Rapids, Michigan, nationally.
1999:	The name "Second Harvest" changed to "America's Second Harvest." Retail store donation program started.
2000:	America's Second Harvest merges with FoodChain. Increasing national focus of America's Second Harvest on obtaining fresh produce.
2001:	America's Second Harvest creates National Advisory Council. America's Second Harvest aids in the 9/11 disaster response.
2002:	First national "Hunger Awareness Day."
2003:	America's Second Harvest starts school pantries in various parts of the country, which become a national program in 2009.
2004:	America's Second Harvest name changed to "America's Second Harvest—The Nation's Food Bank Network."
2005:	John van Hengel passes away at 82, leaving a global legacy.
2006:	Global Food Banking Network formed.
2008:	America's Second Harvest changes its name to "Feeding America." Feeding America focuses on expansion of outreach programs to help connect people to the USDA's Supplemental Nutritional Assistance Program (SNAP).
2011:	Map the Meal Gap research conducted.
2012–Present:	

Many innovations continue:
- Job training programs – operations culinary, truck driving, logistics programs
- Root cause focus – working further upstream of why people are food insecure

- Health and Hunger – development of collaborative nutrition programs
- Neighbor's Voice – advocacy by the people who are food insecure
- Service insights – developing a deeper understanding of the people served
- Meal Connect – the use of technology for more efficient food distribution

ABOUT THE AUTHORS

Dave Krepcho dedicated 34 years to nonprofit sector leadership, led and worked with teams to distribute over one billion meals, generated hundreds of millions of dollars for hunger relief, engaged thousands of volunteers, and created innovative programs and social entrepreneurial initiatives. Dave has served on dozens of local, state, national, and international boards. He has chaired various national task forces, won numerous awards, and testified to Congress on hunger relief. In addition, Dave coaches small nonprofit organizations with the goal of greater mission impact. He believes that food is a fundamental human right and that no one should go hungry, especially in one of the world's wealthiest countries.

Dave lives in Florida with his wife Lois, family, and Quinn, a very protective schnauzer.

davekrepcho.com
krepcho@gmail.com

Claire Strom is the Rapetti-Trunzo Chair of History at Rollins College. She has published widely on global, national, and local history. For this project, she drew on her expertise in agricultural history, including nearly two decades as editor of the journal of note in the field, Agricultural History.

Journey Institute Press

Journey Institute Press is a non-profit publishing house created by authors to flip the publishing model for new authors. Created with intention and purpose to provide the highest quality publishing resources available to authors whose stories might otherwise not be told.

JI Press focusses on women, BIPOC, and LGBTQ+ authors without regard to the genre of their work.

As a Publishing House, our goal is to create a supportive, nurturing, and encouraging environment that puts the author above the publisher in the publishing model.

Storytellers Publishing is an Imprint of Journey Institute Press, a division of 50 in 52 Journey, Inc.

www.ingramcontent.com/pod-product-compliance
Lightning Source LLC
Chambersburg PA
CBHW020448130626
46549CB00001B/342